MOTORING THE FUTURE

MOTORING THE FUTURE

VW and Toyota Vying for Pole Position

Engelbert Wimmer
With contributions by Arun Mani

PA Consulting Group responsible for the content and owner of the IP

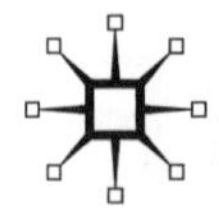

Translated by Brian Melican

First published 2012 by
PALGRAVE MACMILLAN

Palgrave Macmillan in the UK is an imprint of Macmillan Publishers Limited, registered in England, company number 785998, of Houndmills, Basingstoke, Hampshire RG21 6XS.

Palgrave Macmillan in the US is a division of St Martin's Press LLC, 175 Fifth Avenue, New York, NY 10010.

Palgrave Macmillan is the global academic imprint of the above companies and has companies and representatives throughout the world.

Palgrave® and Macmillan® are registered trademarks in the United States, the United Kingdom, Europe and other countries.

ISBN: 978–0–230–29955–9

This book is printed on paper suitable for recycling and made from fully managed and sustained forest sources. Logging, pulping and manufacturing processes are expected to conform to the environmental regulations of the country of origin.

A catalogue record for this book is available from the British Library.

A catalog record for this book is available from the Library of Congress.

10 9 8 7 6 5 4 3 2 1
21 20 19 18 17 16 15 14 13 12

Printed and bound in Great Britain by
CPI Antony Rowe, Chippenham and Eastbourne

CONTENTS

List of Figures and Tables vii
Acknowledgments ix
Preface xi
About the Author xiii

1 The Car Industry Today: Staging a Stellar Comeback 1

2 Volkswagen: The Surprise Challenger 26
- 2.1 Shaping a Global Automotive Empire 27
- 2.2 Strategy for Reaching the Top 33
- 2.3 Showdown in the USA 41
- 2.4 Powerful Competitors 46
- 2.5 China's State-Controlled Economic Boom 47
- 2.6 Expansion in Emerging Economies 65
- 2.7 VW Pushes Sales 66
- 2.8 Prefab Variety 68
- 2.9 Growing without Getting Bigger 72
- 2.10 Intelligent Brand Strategies 75
- 2.11 Process Is the Problem 78
- 2.12 Going the "Volkswagen Way" 81
- 2.13 Small Costs Big 84
- 2.14 Luxury Becomes Obligatory 89
- 2.15 The Search for the Right VW Engine 90
- 2.16 The Price of Politics 104
- 2.17 Piëch, Then Winterkorn – and Who Next? 113

3 Toyota: An Auto Giant Overcoming a Gigantic Crisis 120
- 3.1 The "Toyota Shock" 123
- 3.2 A New Beginning 131
- 3.3 The Power of Tradition: How a Japanese Loom Builder Became an Automotive Giant 134

3.4 The Legend and the Reality Behind It: The Toyota Production System 140
3.5 A Global Player with a Distinctly Japanese Identity 152
3.6 Toyota's Luxury Strategy: Lexus vs. Daimler and BMW 158
3.7 Motorizing Emerging Economies: Toyota's "Car for the World" 162
3.8 Managing Relationships the Japanese Way 165
3.9 Toyota: An Endurance Runner 172
3.10 The Disaster: "Runaway Toyotas" 177
3.11 Change of Generations: Exploring Future Possibilities and Carrying the Load of the Past 190
3.12 Environmental Protection: Toyota's Development of Alternative Drivetrains 193

4 The Race for Pole Position: Winning by a Nose 210
4.1 Phoenix from the Ashes 210
4.2 Round Two 212
4.3 Volkswagen Moves to Overtake 214
4.4 Toyota: Digging in on the Green Front 217
4.5 The Battle on the Bottom Line 218
4.6 The Tides of Globalization 222
4.7 The Final Lap 225
4.8 Comparing Corporate Values 226

5 Epilogue: New Challenges and New Engines 228
5.1 Population Growth and Oil Supplies 228
5.2 Biofuels: Green Gasoline 231
5.3 A Tank Full of Gas 236
5.4 Hoping for Hydrogen 236
5.5 Electric Cars: The Day *after* Tomorrow 237

Notes 255
References 260
Index 265

LIST OF FIGURES AND TABLES

FIGURES

1.1 Global sales forecast until 2015 by region 14
1.2 Car segments transforming 20
2.1 Sales forecast for China through 2015 50
2.2 Transition from conventional platforms to modular toolkits 69
2.3 Estimates of average carbon emissions of fleets from various European and American manufacturers 91
2.4 Total cost of ownership scenarios for conventional, hybrid, and electric vehicles in USA and Europe 93
2.5 Calculating the costs of lowering CO_2 emissions through 2015 98
3.1 The Toyota concern: a network of companies 166
3.2 Supply management: Japanese versus European/US model 169
3.3 Sales forecast for the USA until 2015 177
3.4 J.D. Power and Associates – 2011 vehicle dependability study, US market 189
4.1 The appendix of corporate values 227
5.1 Global population growth 229
5.2 Worldwide oil consumption 230
5.3 Oil price curve based on yearly averages 231
5.4 Alternative fuels: an overview 232
5.5 Technology hype cycle 241
5.6 Fuel usage distribution in an average vehicle – where does the gas go? 243
5.7 Changes in battery costs per kwh 244
5.8 Global sales forecasts for plug-in-hybrid and electric vehicles 252

TABLES

1.1 Comparison of 2010 sales and revenue 13
1.2 Analysis of segment development 21
2.1 Brands and acquisitions of Volkswagen AG 40
2.2 Car sales of the ten largest OEMs in China, January–July 2009 compared to January–July 2010 53
3.1 Major shareholders 173

ACKNOWLEDGMENTS

The work at hand represents an extended and thoroughly revised edition of *Antrieb für die Zukunft*, its German-language predecessor. In updating this book for translation, I have incorporated new facts and material, extending the scope to the American manufacturers, who have staged a stellar comeback in 2010 and 2011. I also took a closer look at new competitors from the Far East, especially the up-and-coming Korean Hyundai/Kia group, inasmuch as they too have taken up the challenge of racing to the top in terms of volume growth and next-generation vehicles.

Antrieb für die Zukunft was written with two co-authors: Petra Blum and Mark Christian Schneider. This new edition would not have been possible without their profound input for the German book, which was published in 2010.

Furthermore, I remain very grateful to the two companies so closely portrayed in this book. Both Volkswagen and Toyota were very supportive of the project, and I would like to express my deep gratitude to the managers at both companies for the many insightful conversations we shared on the topic.

I would also like to mention Arun Mani, an energy expert and member of the PA Consulting Group management team. Arun made a decisive contribution to the last chapter with his views on the interplay of the energy sector and the car industry (section on "Utilities and the adoption of electric vehicles" in Chapter 5). I am greatly indebted to him for his help in showing where the frontiers will lie in the future of automotive mobility.

In fact, there was a whole team of people at PA Consulting who worked on this book, and I would like to extend my thanks to Karsten Gross from Marketing, consultants Felix Salditt and Michael Tickle, and Thorsten Brückner from the PA Knowledge Processing Centre.

Deserving of a very special thank you is the ever-accurate Hanns Peter Becker, a consultant and engine specialist who was indispensible in making our mathematical models work. I would also like to thank Brian Melican, a freelance writer, journalist and translator, for his patience, endurance and hard work in putting this work into English and copy editing the final stage.

Finally, and most importantly, I would like to thank my family for their unfailing support.

PREFACE

The story of this book begins back in 2007, when Martin Winterkorn, having just taken office as CEO of the automaker Volkswagen, called out the worldwide market leader Toyota. By 2018, Winterkorn declared, the German Volkswagen Group would surpass the Japanese industry champion and become the number one automotive manufacturer in the world. At the time, the race between the then comparatively much weaker VW group and the seemingly all-powerful Toyota Motor Corporation seemed more than unequal. Yet the idea that VW was trying to overtake Toyota, as improbable as it then seemed, thrilled the entire industry – and, not least, inspired the author of this book to discuss the topic with industry experts and automotive managers all over the world. After many interesting discussions and a lot of research, the idea was born to distill all this knowledge into a narrative.

The first edition of this book was published in 2010 and told the story of how VW would go about climbing to the top. Taking the very public challenge to Toyota as a narrative backdrop, the first edition of *Motoring the Future* analyzed the strengths and weaknesses, corporate cultures and philosophies of these two so very different industrial icons. The analysis often produced surprising insights, not only about these two carmakers, but about the automotive industry itself.

Since 2007, the race between VW and Toyota has taken an unexpected turn. Global economic and financial crises, bankruptcies and resurrections of major rivals, governmental intervention and a natural catastrophe which sparked a nuclear disaster in Japan have quickened VW's advance, while Toyota has had to overcome one crisis after another; ironically, the event which has the least long-term importance for Toyota – the earthquake and Fukushima meltdowns – will probably be the decisive one as far as VW's avowed goal is concerned.

2011 may well now turn out to be the year in which Volkswagen overtakes Toyota; but the consequences of the natural catastrophe that struck Japan, destroying factories and disrupting supply chains, are not permanent. Toyota has fallen behind unexpectedly, but this will, if anything, spur it on to redouble its efforts in coming years. The race is by no means over. Moreover, the rivalry depicted in *Motoring the Future* offers a deep insight into the challenges that are looming on the horizon for carmakers: as they have to move away from fossil-fueled combustion to alternative energy vehicles for the mass market, the economic environment for car makers will change tremendously. Each of the major players in the automotive industry is trying to answer the $64,000 question: just how will the car of the future look? This is the decisive field on which the race for control of the industry will be run, and the two competitors – VW and Toyota – will play a vital role in leading the industry into the new era of green vehicles.

Since the first edition of *Motoring the Future*, I have analyzed and updated Volkswagen's and Toyota's strategies for next-generation vehicles; both are breaking new ground in this area, plotting a course into the future for other manufacturers and suppliers. Whether it is the Japanese or the German carmaker that takes pole position, the winner will change the balance of power in the worldwide automotive industry and show precisely which technologies, philosophies and strategies are needed for car manufacturers to be successful in the market of the future.

ABOUT THE AUTHOR

Engelbert Wimmer is a business consultant and member of the management team at PA Consulting Group, a leading management and IT consulting and technology firm. Based in Frankfurt, Engelbert leads the firm's global manufacturing and automotive practice. His main focus lies on projects at the critical interface between scientific innovation and business.

With his experienced team, Engelbert has helped global manufacturers and suppliers alike on cost and performance issues. He is an expert on the car industry and frequent speaker at automotive industry events.

1

THE CAR INDUSTRY TODAY: STAGING A STELLAR COMEBACK

Car manufacturers used to have to be able to see 15 years into the future. Nowadays, a whole automotive era goes by in a year.

— Osamu Suzuki, head of the eponymous Suzuki Group, speaking in December 2009 as Volkswagen took a stake in his company

On May 23, 2011, New York's world-famous Museum of Modern Art was open slightly later than on an average Monday. Not for the general public, of course, but until well into the small hours, a coterie of A-list stars including Yoko Ono and Madonna could be found exchanging pleasantries with leading Volkswagen executives in the ultra-modern surroundings of the MoMA. The occasion for this high-class dinner event was the announcement of a new partnership between the museum and the German automaker, with Volkswagen pumping several million dollars into MoMA as part of a PR offensive which was supposed to ramp up Volkswagen's presence on the US market.

Presenting representatives of the museum with "Tornado" and "Mirage", two video installations by the Belgian video artist Francis Alÿs, Volkswagen CEO Martin Winterkorn waxed lyrical over the connection between MoMA and Volkswagen as "world-wide brands".

Despite his warm words however – and the fact that his birthday fell on the following day – Winterkorn did not stay at the party for long. Indeed, while many of the guests were still strutting their stuff on the dance-floor, the man at the head of the Volkswagen Group was sound asleep in his hotel room. He hadn't even stayed up long enough to toast the start of his 65th year with his entourage.

Not that there was no reason to celebrate - but Martin Winterkorn's tight schedule did not leave time to make a night of it. Just a few hours later,

Winterkorn was several thousand feet above the Eastern Seaboard in the VW company airplane on his way to a festive event a hundred miles or so south of Nashville, TN.

The reason one of Germany's premier captains of industry ended up spending his birthday in Chattanooga, this small city of 167,000 inhabitants known as the "Gateway to the Deep South", has nothing do with Southern hospitality or any special appreciation for Glen Miller and his classic "Chattanooga Choo Choo".

Chattanooga is important to Winterkorn, but the attachment is still more due to business than to individual infatuation: the Volkswagen Group has big plans for this small city. By 2018, the German carmaker wants to be nothing less than the biggest automotive manufacturer on the planet, and Winterkorn is convinced that anyone who wants to reach that size needs a strong foothold in the USA. That is why Volkswagen is opening a new factory here, staffed by over 2000 workers and creating almost 10,000 further jobs at local suppliers.

Suddenly, this average town on the gentle, picturesque loops of the Tennessee River is a big name several thousand miles across the Atlantic in Wolfsburg, the "car city" that Volkswagen calls home; and on May 24, 2011, almost everyone who is anyone back at German HQ was in Chattanooga with Winterkorn.

Volkswagen has a lot of catching up to do on the US market, and a lot of ghosts to lay to rest. The low point was 23 years ago when, following years of loss-making production, the Rabbit assembly plant in Westmoreland, PA was closed. It was the first and only time that the automaker has been forced to close a site due to poor sales; even today, Volkswagen executives shudder at the memory of this traumatic event.

In retrospect, there were plenty of obvious reasons why the plant at New Stanton, Westmoreland County did so poorly. The Rabbit model, known as the Golf in Europe, was the wrong product for the American market, too expensive by half and assembled in a shoddy fashion by poorly-motivated employees who spent more time checking UAW regulations about working hours than carrying out quality checks.

Volkswagen has always been an ambitious company, and this ambition led them to blindly sink billions of dollars into the USA, the land of the automobile and of big business: it is this same ambition that means that, despite the scale of their defeat there, Volkswagen will not stay away from the American market. As far as the German Group is concerned, USA is unfinished business, and the Chattanooga plant is the base from which it intends to re-conquer the American market.

On the way from the airport to plant 62 – built, fitted and ready to enter service within just two years – the convoy of German executives drove by billboards with their local slogan "Lives here. Works here. Volkswagen"; the adverts were placed along the newly-renamed Volkswagen Drive.

This is part of Winterkorn's strategy to appeal to American national pride, with the message being that VW is "at home in America like never before." At the inauguration ceremony, Winterkorn took the opportunity to expand on this: "The Volkswagen Group has finally arrived as a local manufacturer in the United States. We are proud to be part of this great automobile nation as a producer, an employer and a friend and good neighbor to people in the region."

Then again, the US factory will not just bring economic benefits to South Eastern Tennessee, but to Volkswagen too. Until recently, all VW models except the Jetta (built in Mexico) had to be imported from Europe, resulting in high losses: Winterkorn is counting on the new plant to change this.

The German automaker is willing to spend money to make money, with investments in the Chattanooga plant totaling roughly US$1 billion (€700 million). Further to this, Volkswagen has spent a lot of money developing a considerably larger, 15-foot version of the European Passat and a new engine assembly plant in Mexico. Once all of this is taken into account, VW has spent more like US$3 billion (€2 billion) on this American venture. Then again, they stand to profit from at least half a billion dollars in state concessions and subsidies over the coming years.

In fact, it would seem that Volkswagen's huge investment program is coming on stream at just the right time. At the Chattanooga inauguration, US Secretary of Transportation Ray LaHood sent a signal about future federal policy that European car manufacturers have been waiting decades to hear. "We can build a 21st century transportation system", he promised whilst praising VW's clean-diesel engine model. Winterkorn was quoted as saying that this was the best birthday present of all.

For years now, German carmakers have been fighting to roll out diesel-driven engines in the US market to the same extent as they have in Europe; but their success has been limited. Yet now there are different noises coming from Washington, and the German car industry is confident that America is starting to discover the advantages of fuel-efficient clean-diesel technology – promising even higher sales and handsome profits for diesel-focussed German carmakers – and above all, Volkswagen.

In the industry, there is a buzz around Volkswagen like there hasn't been for decades. It is Europe's number-one car manufacturer, the world's confirmed second largest on revenue (and quite possibly on sales), and not only has it survived the lean years: it has gotten stronger. VW is clearly on its way to the top. Fuel-efficient technology, a new, ultra-modern production site in the States, a broadly based PR offensive and a new model, the mid-sized Passat, competitively priced with less than US$20,000 on the windshield – Volkswagen has equipped itself with what it thinks it takes to beat its competitors in the US market. In order to conquer North America, however, Volkswagen needs to succeed over Toyota – the industry giant it called out several years ago.

With Toyota, Volkswagen has set its sights on the pioneer in fuel-efficient hybrid technology, the manufacturer of the most-sold car in the world, the Corolla, and the carmaker which for many years dominated the US market, not only in the compact, but also in the luxury segment.

The USA was the place where the somewhat unspectacular manufacturer from Toyota City had made it into the big leagues, had become a global player, conquering the premium market with the Lexus brand and polishing its image around the world as the leader in environmental-friendly technology with the mass-market hybrid car, the Prius.

In 2009 and 2010 however, North America became the site of Toyota's most ignominious defeat.

The North American market was where Toyota piled up huge losses during the 2008–2009 crisis, being greatly responsible for red figures the company carried in its books in 2009 (the Japanese fiscal year, which Toyota applies, begins April 1 and ends March 31). In just the three months between January 1 and March 31, 2009, the carmaker had lost more than ¥682 billion (US$7.3 billion / €5 billion), making for a worse operating result than General Motors, which at the time was in the process of filing for Chapter 11. It had taken little more than 90 days to turn the car company that produces more vehicles than any other in the world into the car company that was losing more money than any other.

Even worse, since fall 2009, Toyota has had to recall some 11 million vehicles – most of them in USA – for faulty gas pedals, floor mats that could trap accelerators and braking and engine defects, each recall tarnishing the company's once-sterling reputation for quality. Akio Toyoda, grandson of the company's founder Kichiiro Toyoda, had to testify before the American Congress, a scene of the courtroom sort unimaginable to a Japanese industry captain. As a result of the quality issues, Toyota's market share in USA has decreased, despite generous incentives. Just four years ago, Toyota had knocked Detroit rival Ford from the number two spot – in 2010 and 2011, however, Ford was able to reclaim its position as number two American carmaker after GM – and ahead of Toyota.

Then, after two grim years in North America, another crisis hit the giant, this time in its home country.

On March 11, the Great East Japan Earthquake and ensuing Tsunami shook Japan to the core. The island nation experienced a natural disaster and subsequent nuclear catastrophe of unprecedented dimensions, and the horrible scenes of devastation emerging from the regions affected struck at the hearts of everyone in the nation. For months after the earthquake, Toyota's CEO

Akio Toyoda spoke of little else except his condolences to all victims of the disaster and offered prayers for those who lost their lives in the earthquake and the tsunami.

Japan's car industry suffered badly from the effects of the natural disaster. Toyota's production sites were idle for months because supply chains were interrupted; even in the USA and Europe, assembly lines had to be stopped. Japanese carmakers lost at least a million sales as a direct result, and Toyota was hit hardest; Akio Toyoda had to go on record saying that production would not be back to normal before November 2011.

If Volkswagen does overtake Toyota as early as 2011, it will be in no small part due to the earthquake and tsunami. Asked about their chosen rival's difficulties, Volkswagen's reaction was modest and respectful. VW did not want to be number one because of a devastating natural disaster, VW executives said.

As far as 2011 is concerned, Toyota has fallen behind. The giant took a full broadside and will need time to lick its wounds, but the Japanese carmaker is down, not out. Quite to the contrary, say most industry experts, who readily observe that only a giant with the strength of Toyota could digest such a merciless succession of crises.

As a consequence, in the short term, it is quite possibly down to another competitor to prevent Volkswagen from climbing to the top spot, a rival no one would have expected to recover from the lean years so quickly and so strongly. Volkswagen's new rival is an American carmaker which seemed to have fallen prey to the most dangerous period of turbulence in the automotive industry since the oil price shocks of the 1970s – above all in USA, where the economic crisis hit hardest and fastest.

It was from here that the crisis had spread, as early as October 2008. Mere weeks after the collapse of the New York-based investment bank Lehman Brothers, car sales – already slowed by high gas prices – ground to a near-halt as the Wall Street meltdown scared consumers and cut off many from credit. From North America, the reluctance to buy spread to Europe and had reached almost all the major car markets before the year was out. Dire new forecasts for global vehicle sales battered global auto companies and forced them to put the emergency brakes on their assembly lines; pictures of brand-new and unsold vehicles literally piling up in the major global harbors became one of the symbols of the spiraling recession, right up there along with bankers carrying their office effects down Wall Street in cardboard boxes. A wave of insolvency washed over dealers and suppliers; even the lending arms of car companies, one of the most profitable branches for many automotive manufacturers, started putting together applications for state aid. The entire sector seemed to have been plunged into an abyss of chaos within the space of just a few weeks.

After Wall Street and the US banking sector, the US car industry became the second major branch of industry to undergo landslide changes. The kick-off was the never-to-be-forgotten November 19, 2008, when the heads of the Big Three went before Congress to ask for $25 billion (€17.3 billion) in direct aid, of which $10–15 billion were supposed to go to General Motors alone.

Despite heavy criticism, former President George W. Bush announced an emergency bailout of GM and Chrysler, pumping loans of $13.4 billion (€9.3 billion) into the two carmakers. What hardly anybody would have openly admitted at the time was that billions more would have to follow. Nevertheless, from that day on, the American carmakers were set on a course for a merciless restructuring and, following this, an astonishing renaissance.

General Motors, for decades the largest automaker in the world based on sales volume, was forced to file for Chapter 11 before deliberately shrinking down as almost half of the Group's brands were sold off or shut down. Saab went to Dutch sports-carmaker Spyker, while Hummer was closed after the planned sale to a Chinese company fell through. Saturn and Pontiac followed suit: examples of genuine Americana confined to the junkyard of history.

The disaster, however, had not come about overnight. Even before the financial crisis, General Motors had been suffering. In 2007, GM booked a turnover of US$181.1 billion (€125.8 billion) with US$38.7 billion (€26.9 billion) in losses – the biggest deficit ever in the history of the carmaker. In its balance sheet that year, the company had liabilities of US$185 billion against assets of just US$148.8 billion, leaving the company US$37 billion in excess debt. Market share had been deteriorating continuously: in 2000, GM still had sold 4.9 million vehicles in the USA; eight years later, this figure had more than halved to just over 2 million.

In the fall of 2008, despite two years of steep cutbacks, GM found itself on the brink. In March 2009, President Obama forced out GM's chief executive, Rick Wagoner, rejected the company's restructuring plan and forced it into bankruptcy after its creditors had balked at deep write-downs.

On June 1, 2009, General Motors, by this date more than 100 years old, filed under Chapter 11. Further examination of the balance sheets revealed that by this point GM had just US$82.3 billion to its name against colossal debts of US$172.8 billion.

In announcing the bankruptcy, President Obama envisioned that a much smaller, retooled GM could make money even if new-car sales in USA remained at a sluggish 10 million a year, and even if the former giant of the industry dropped further below its remaining market share of less than 20 percent.

But to get there, American taxpayers had to invest an additional US$30 billion (€20.8 billion) in the company, atop US$20 billion already spent just to keep it solvent as the company bled cash as quickly as Washington could inject it.

The application was rushed through the courts at record speed, closing on July 10, 2009. GM sold its good assets to a new, government-owned company. Brands like Chevrolet, Cadillac and GMC were folded into the new entity, renamed the General Motors Company, and facetiously referred to as "Government Motors".

Of the new GM, the federal government held nearly 61 percent, with the Canadian government, a health care trust for the United Auto Workers union and bondholders owning the balance.

The restructuring was brutal. Hourly labor costs were cut by more than two-thirds, from US$16 billion to 5 billion in 2005. The insolvency saw GM reduce its manufacturing plants from 46 in early 2008 to merely 37. Two thousand six hundred of GM's American dealers were eliminated, making up for 40 percent of the entire dealership. The cuts were meant to thin bloated dealer ranks that were a holdover from the company's better days.

Meanwhile, in Europe, the carmaker struggled to secure state funds for its subsidiary Opel/Vauxhall before renouncing its claims eventually without success and after a truly nerve-wracking period of negotiation. The number of employees on GM's books fell from almost 270,000 before Chapter 11 to roughly 200,000 after the restructuring.

How did the company come to this? For most of the twentieth century, General Motors had been the biggest company in one of the most important industries in the world. It not only led in automotive innovations, but helped define the new breed of massive, bureaucratic multinational corporations that shaped the US post-war economy. It was the world's largest carmaker from 1931 to 2008, when it was outgrown by Toyota. It seemed that it had lost its feel for reading the American car market it helped create, while Japanese automakers lured away even its most loyal buyers. Foreign competition, and GM's failures to provide trustworthy cars that Americans wanted, began to diminish its luster and its sales – as had happened with its two main American rivals.

The Big Three from Detroit: Rise, Fall and Rise again

In 1988, 74 of every 100 cars sold in the USA carried the insignia of one of the big three American automotive manufacturers: Ford, General Motors or Chrysler. By 2008 this figure had sunk to 48 percent. Trivial design, quality issues and poor comfort: these were the reasons why American customers started to desert their domestic carmakers in waves. Asian carmakers' share of the US market, meanwhile, climbed by 21 percentage points to 45 percent, 17 of which went to Toyota alone.

Making a bad situation worse, Detroit's Big Three spent the 1990s running a very short-sighted shareholder-value strategy which starved investment into new engines and strategic compact and subcompact cars. Furthermore, the Michigan manufacturers were laboring under the burden of high production costs, and in the growing compact segment above all they were completely unable to catch Toyota and its lean production methods in profitability. The only vehicles which still turned a profit were SUVs and heavy goods vehicles.

In order to cut costs, the large American manufacturers struck deals with the unions in the mid-1990s: the workers agreed not to demand pay increases and to refrain from striking; in return, they were accorded long-term pension plans and healthcare benefits that were far above the average settlements for other domestic manufacturing sectors.

Some of these benefits were not agreed just for the workers themselves, but for family members or workers who were already claiming their pensions. The automotive majors set up funds in order to cover such costs, but these were chronically underfunded for years and led to poor balance sheets for GM, Ford and Chrysler. At the same time, the aging workforce in Detroit sent healthcare costs through the roof. Former Governor of Massachusetts Mitt Romney famously argued in *The New York Times* that the high wages, pensions, and healthcare costs at the Big Three were adding US$2000 to the costs of every vehicle they sold. According to Romney, US$2000 was the amount the American manufacturers were forced to save in terms of quality just to compete on price in the face of the growing competition from foreign manufacturers.

These huge insurance liabilities pressed down on bottom lines like lead weights in the balance sheets of Detroit's Big Three, cutting their leeway for innovative entrepreneurial thinking and weakening them even further against the Japanese competition.

The high proportion of in-house production practiced by GM, Chrysler, and Ford was associated with high costs, so huge sections of these companies had to be sold off and set up as independent suppliers. Just to keep the core business above water, General Motors was forced to shed American Axle in 1994 and the gigantic supply operation Delphi in 1999; Ford, likewise, had to split off Visteon in 2000. This led to thousands of workers being "outsourced" to the new companies, who were not slow to cut their employment costs: by 2009, the American car industry (including suppliers) employed 880,000 people, roughly half a million fewer than in 1999.

Yet costs were only half the problem for the Americans. In comparison to competitors, the Big Three were slow to recognize that rising gasoline prices would turn small compacts into a growth segment and were reluctant to move from big high-margin vehicles to building small, less profitable cars, even when gas prices began a steady rise after 2004. In the end, they were left scrambling to fill the strategic gap, and seldom made any profit doing so. They often had to sell ten or more compacts to reach the profit they made from a single SUV, and in order to keep their expensive plants running and stop their market share from falling they were forced to sell *en masse* at discount prices. Hundreds of thousands of vehicles a year were pushed into the market at dumping prices (mostly through fleets and leasing), only to end up cheap on the second-hand market. As a result, uncompetitive vehicles came in heaps onto the US market and pushed the price level right down.

The US car market was even more strongly affected by the 2008–9 financial crisis than almost every other worldwide; SUVs and light commercial trucks sat on dealership lots gathering dust. American automotive manufacturers had to dramatically scale back production, and it looked as if after a long agony the US car industry was in its final death throes. In November 2008, the CEOs of the American manufacturers were forced to go before Congress and, cap in hand, publicly request inordinate amounts of government money.

On the morning of November 18, 2010, two years after the memorable and humiliating appearance of the former heads of the Big Three before Congress, new GM chairman Daniel Akerson arrived at Wall Street before the market opened to ring the bell that signals the beginning of the trading day – to many, the official wake-up call for a legend and the American car industry as a whole, two years after a near-death experience. GM later said that it had been accompanied by the sound of a 2011 Chevy Camaro SS engine revving. A whole procession of vehicles – from the GM stables – filled Wall Street. On the front of the Stock Exchange a giant blue GM sign reading "GM NYSE Listed" was posted with a new, silver logo that replaced GM's familiar, square insignia. One and a half years after one of the biggest corporate collapses in history, GM launched a historic public offering, the biggest ever seen in the USA. General Motors had made one of the biggest rebounds ever.

"Americans love a comeback story. We're going to play like underdogs," Akerson later told the media – and a comeback story it was, indeed one of the very American kind. After American taxpayers had put up US$50 billion to bail out GM, an IPO nearly halved government holdings to 33 percent, as GM raised US$15.77 billion, as well as US$4.35 billion in preferred shares.

With the success of the IPO, there is no more doubt about GM being on the mend. Whilst its decline was slow and painful, its resurgence has come surprisingly quickly. Out of the remains of the lumbering, tired old giant has sprung a lean, agile, and hungry tiger, and sales figures in USA have started to rise again. GM retained the top spot with US sales of 2.2 million vehicles in 2010, which made a US-market share of 19.1 percent. The company earned US$4.7 billion in 2010, its first full-year profit since 2004.

Remarkably enough, sales in China have become a crucial part of the carmaker's long-term strategy. In 2010, its China sales stood at 2,351,610 units, compared to 1,826,475 units in 2009, an increase of 28.8 percent. For the first time ever, the automaker sold more vehicles in China than in the United States. Just 13 years after entering the People's Republic, GM now says the country accounts for a quarter of its global sales – blistering growth that even in Detroit hardly anyone expected this soon. In mid-2011, however, GM saw weakening sales in China, as the government began withdrawing stimulus measures such as tax breaks for small engine vehicles which had been aimed at cushioning the impact of the global economic downturn.

So far, American taxpayers are still losing on the bail-out deal: for taxpayers to break even, GM shares would have to trade at about US$53 (€36.8) a share. Only few analysts predicted this kind of price range for GM stock at launch; meanwhile, in mid-2011, with much of the initial excitement gone, and the continuing recession risks in the US economy, the stock is performing poorly.

It also still remains to be seen if the revived GM is positioned to regain lost market share in the strategically important compact and green car segments.

GM's flagship electric car, the Chevy Volt, meanwhile is on sale with US$39,995 on the windshield. That price comes down, however, as electric car buyers in USA can apply for a federal tax credit that ranges between US$2500 and 7500, depending on the size of the battery in the car. On the low end of the spectrum, cars with 4 kWh battery packs qualify for a US$2500 tax credit. The Chevy Volt, in contrast, maxes out at US$7500 because of its 16 kWh battery pack. Customers can even save more money in states such as California, Louisiana, Hawaii, Oregon or Colorado, which have their own robust EV incentive programs in place.

Moreover, GM brought the car's price closer to its rival, the pure plug-in electric Nissan Leaf. Sales of both electric cars have been neck-and-neck in the first half of 2011. As of the end of May 2011, Nissan had sold 2184 Leafs while GM had sold 2167 Volts. As EV sales were not picking up at the expected rate and the competition for the few electric car enthusiasts is intense, it seems that the price war for this market has already begun, before the market has taken off.

Moreover, in fall 2010, GM rolled out one the most significant new-model introductions in the United States since its 2009 bankruptcy, the Chevrolet Cruze. It is the most earnest attempt by any of the US Big Three (GM, Ford and Chrysler) to build a compact car that Americans might buy because they like it – not simply because it is cheap. The Chevrolet Cruze even led compact car sales in May 2011 – for the first time in five years, an American car was the best-selling compact car in the United States, a segment that had been dominated by cars such as the Honda Civic or the Toyota Corolla. Until mid-2010, winning over American consumers with a small, fuel-efficient car, long the domain of Japanese or Korean carmakers, had mostly eluded US automakers – only Ford stood out among its competitors.

Ford – another stellar comeback

After the start of the millennium, Ford – the company which invented modern automotive mass production – had been in constant danger of falling into utter irrelevance. Although in 2000 US sales figures were still at 4.2 million, by 2009 Ford sales in the USA had sunk to 1.7 million.

While the sales decline had been similar to that of GM and Chrysler, however, Ford has been spared a trip to the bankruptcy court in the wake of the economic crisis as a result of radical cuts pushed through by management in 2005–6. Implemented under the moniker "The Way Forward", from January 2006 onwards the plan saw Ford adapt itself to the new market conditions in several ways. Unprofitable models were discontinued, several production lines and plants shut down. The car-rental subsidiary Hertz and shares in the Japanese manufacturer Mazda were sold along with the brands Aston Martin, Jaguar, Land Rover and Volvo. Job losses at Ford were in five figures.

Ford's new CEO, Alan Mulally, who came from Boeing, got behind this new direction and pushed the plans through early and aggressively, allowing Ford to cope comparatively well during the economic crisis. Moreover, Mulally had the engineers produce more economic cars whilst he also cut the high healthcare and pension bills. As the crisis unfolded in 2008, Ford was able to persuade its creditors to trade in the debt against shares in Ford. In this manner, the company was able to restructure US$10 billion (€6.9 billion) of its mountainous debt while saving almost US$500 million (€350 million) a year in interest payments.

In the US domestic market above all, car buyers have been eager to pay tribute to Ford's strength and strong desire to stay solvent without taxpayers' money. Having understood earlier than other American manufacturers that smaller, fuel-efficient cars would be a growth area in USA, Ford has added competitive edge. The Ford Focus, for example, a compact developed in Europe, has blossomed into a sales hit in the USA, profiting to no small extent from the "Cash for Clunkers" sales incentive scheme which had been initiated by the US government 2009.

Moreover, Ford's turnaround has been gaining momentum as it has increased sales and market share in the USA. Full-year profits for 2010 climbed to US$6.6 billion from US$2.7 billion in 2009, the highest annual income in more than a decade. Not that Ford has by any means finished restructuring: the company must continue to work on its balance sheet, paying down debt as it generates operating cash. Ford still faces challenges, such as the automaker's global pension funds, which were reported to be US$12 billion (€8.3 billion) underfunded at the end of 2009. Analysts estimated that funded status could deteriorate a further US$5.5 billion, to about US$17.5 billion. Pension expenses might turn out to be the bad news in an otherwise remarkable success story.

Conversely Detroit's number three, Chrysler, widely perceived to be the weakest and most troubled of the domestic Big Three carmakers, turned out to be the next success story. After it going to a group of investors, then through Chapter 11, and finally to the Fiat Group, (the pension funds of the UAW, and the governments of USA and Canada also hold interests), in late 2010, it was closely eyeing its cross-town rival's IPO success. In May 2011, Chrysler paid US$7.5 billion in government loans back – to both the US and Canadian states. Chrysler had owed US$5.9 billion to USA and US$1.6 billion to Canada, and had until 2017 to repay its debts in full. The loan payoff came far ahead of what was required, and with this financial move, Chrysler hopes to offer public stock later this year or the next, based on market conditions and the wishes of its shareholders.

With Fiat's help, Chrysler has revamped its product lineup, redesigned its vehicles and is now fetching higher prices. In the first quarter of 2011, the carmaker reported its first net profit of US$116 million. Chrysler and Fiat CEO Sergio Marchionne said Chrysler is on track for net revenue of US$55 billion in 2011, a US$2 billion operating profit and net profit of US$200–500 million.

Overall, Chrysler's turnaround strategy is still less advanced than those of its Detroit rivals. In a large part, its revival is based on pairing the Chrysler and Fiat product lines. Chrysler re-labels cars of the Fiat brand Lancia for the US market, in return Fiat uses Chrysler vehicles to extend Lancia's range in Europe. Chrysler still has considerable work to do in revamping its selection of cars and trucks, but it has shown significantly more promise under the control of the Italian automaker than before its bankruptcy, when it was owned by a private equity firm that spent little on product development. In 2011, Fiat raised its stake in Chrysler to more than 50 percent by buying the US Treasury's remaining 6 percent interest in Chrysler, boosting Fiat's stake to 52 percent.

GM's success provides potential lessons for Chrysler in terms of cost cutting, improving the range on offer and communicating successes effectively, although experts often point out that something might be holding Chrysler back in comparison to GM: the lack of presence in the ever-growing Chinese market. The China story has been an important one for GM during the run-up to its IPO. Analysts even went as far as to say that without it, there wouldn't have been much of a deal. Meanwhile Chrysler is exploring whether it can start manufacturing vehicles in China by using a plant its partner Fiat is building in the country and which is supposed to start production in 2012. Using Fiat's capabilities would quickly allow Chrysler to gain a production foothold as it fights to catch up with its rivals.

As growing Chinese sales have spurred GM's business, even going head to head with Toyota in global 2010 sales, at the same time the People's Republic is nurturing Toyota's new rival Volkswagen, which in 2010 has become number three in sales and number two on revenue worldwide – thanks to strong earnings in China, above all (Table 1.1).

Like the USA, with its huge sales potential, China has a special role to play in the carmakers' fight for the top spot. Until recently, the global car industry had seen the People's Republic as little more than a sales receptacle with huge potential. China, however, is growing ever-more confident, trying to rid itself of its low-quality image and even aiming to break into European and US markets (Figure 1.1).

Whenever it appears advantageous to do so, the Chinese also acquire established Western brands whose parent companies have put them up for sale. At the end of March 2010, the largest private Chinese carmaker Geely signed an agreement to purchase Swedish subsidiary Volvo from Ford. The idea in

TABLE 1.1 **Comparison of 2010 sales and revenue**

Manufacturer	Vehicle sales, calendar year 2010, in million	Revenue, fiscal 2010, US$ million (rounded figures)
Toyota (including Hino, Daihatsu and Chinese joint ventures)	8.418	253,000
GM	8.39	135,600
VW	7.14	183,000 (€126,875)
Hyundai/Kia	5.75	revenue figure (36.77 trillion Won)
Daimler	1.895	141,000 (€97,761)
Ford	5.31	121,000

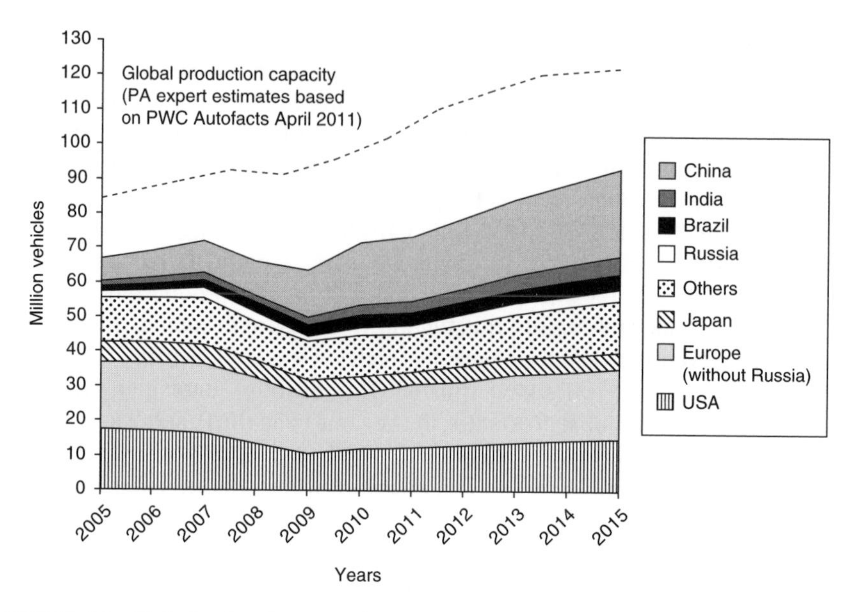

FIGURE 1.1 **Global sales forecast until 2015 by region (all vehicle segments), PA Consult-ing Group.**

Source for historic data: 2005–2010 figures taken from Automotive News Europe (Global Market Data Books) and MarkLines.

Beijing is to piece together automotive conglomerates which could eventually be in a position to call out giants such as Volkswagen and Toyota. The technology would originate from Europe, the USA, and China, while manufacturing would take place in China for the most part. In the hands of Chinese manufacturers Western brands like Volvo are intended to generate flair and a sense of familiarity and trust overseas.

This kind of change in ownership is a symbol for the current rebalancing of power in the global automotive industry. While long-established manufacturers based in the "triad" countries (Europe, the USA, Japan) are struggling with saturated, stagnating markets and fiercer competition, new manufacturers – especially Chinese companies, some of which are wielding considerable financial resources – are muscling their way onto the industry's playing field.

As the crisis has mixed up global car industry and manufacturers are staging their comeback amidst growing global sales, the competitive environment has continued to change, however. The race for the next generation's cars is fiercer and more open than ever. Much of the excitement in major car markets revolves around the question of how a new generation of cleaner cars could be fueled, reducing the car industry's dependence on oil. For society as a whole the challenge of the coming years is how to keep the population mobile without fossil fuels, since oil is getting scarcer or simply more and more expensive to source. This will force all manufacturers to invest huge amounts in research and development; the sums involved will probably be bigger than at any time since the dawn of this industry more than a hundred years ago. Changes affect the automobile's engine, above all, thereby striking at the heart of every car company. The fact is this: cars of all sorts will have to either be driven using far less gasoline and diesel than to date, or run entirely on alternative energy sources.

Political and societal megatrends such as the climate debate are forcing car manufacturers to take account of sinking CO2 emissions allowances. In Europe, 65 percent of new vehicles sold will have to emit 130 grams per kilometer (209.21 g/mi) or below by the beginning of 2012. By 2020, European Union legislation will require all new cars to emit 95 g/km (152.89 g/mi) or less.

USA – federal regulations

Federal regulations concerning vehicle emissions in the USA are organized under the Clean Air Act (CAA) of 1963 (modified in 1970, 1990), which is enforced by the Environmental Protection Agency (EPA). However, until recently the CAA conspicuously failed to list CO2 as a pollutant, leaving the generous Corporate Average Fuel Economy regulations to set standards for CO2 vehicle emissions in the USA. With the American Big Three consistently coming bottom of EPA annual reports on gas mileage, this relatively lax regulation has been a consistent crutch to US manufacturers, whilst European and Asian automakers see little-to-no benefit for their efforts to reduce fuel consumption – and therefore emissions – in past years.

However, in a far-reaching ruling in 2007, the Supreme Court forced EPA to begin taking steps to regulate CO2 emissions from motor vehicles on a national level from the beginning of 2010. Impetus on gasoline consumption also came from the newly elected President Obama, whose administration announced in April 2010 that the Corporate Average Fuel Economy would be raised to 35.5 miles per gallon through 2016; in May 2010, Obama told government agencies to work on a stricter rule to succeed this in 2017.

California

As a response to its chronic air pollution issues, California has consistently been at the forefront of clean air legislation: since it created a regulatory agency for air quality before the first federal legislation on the matter in 1963, it is exempt from EPA control and is permitted to maintain its own regulatory body (CARB – the California Air Resources Board) while other states are only allowed to follow federal lawmaking, or to adopt CARB standards inasmuch as they are at least as protective of public health and welfare as applicable federal standards – and Californian standards invariably are.

In 2002, California passed the first law in the USA to directly address greenhouse gas emissions across the entire fleet. Under the act's provisions, car manufacturers would have had until 2009 to produce "California cars" subject to successive caps on emissions – 22 percent fewer greenhouse gases by 2012, 30 percent fewer by 2016. Despite the fact that the provisions of the law were immediately adopted by seven Northeastern states (known as the "CARB states"), the EPA challenged the legislation on the grounds that it was outside the states' jurisdiction (and its own) to legislate on greenhouse gas emissions, provoking court challenges in turn from the states involved.

However, in a landmark ruling in 2007, the US Supreme Court affirmed that states are legally allowed to regulate greenhouse gases; later in the year, a judge in Vermont sidelined continued objections from automotive companies that California's emissions regulations amounted to a preemption of federal jurisdiction in setting fuel economy, and the state's mandatory limits on motor vehicle emissions are now essentially unchallenged. Further impetus has come in recent years from the low-carbon fuel standard, requiring oil refineries and distributors to successively reduce CO2-equivalent grams per unit of fuel energy. Starting in 2011, the standard requires road fuel producers and importers to establish and report on annual reductions in whole life cycle carbon intensity of fuels. For gasoline and diesel the schedule starts with 0.25% reductions in 2011 and increases in pace to provide a total of a 10 percent reduction by 2020.

Yet California is not just operating with the stick; carrots including the Clean Vehicle Rebate Project are on offer. In 2009, the state of California appropriated over US$10 million to help promote the production and use of zero-emission vehicles, with rebates ranging from US$5000 (€3470) for cars for personal use to US$20,000 (€13,890) for some commercial ZEVs (Zero Emission Vehicles).

Slated to last until the end of 2011, the project distributed about 2000 rebates worth a total of US$11.1 million in just about 27 months. Despite a US$2 million boost from the California Energy Commission, the California Air Resources Board (CARB) in June 2011 announced that the program had run out of money earlier than predicted. The CARB hopes to triple the amount of funding to the program for the 2011–2012 fiscal year to US$15 million which, with reduced award amounts of US$2500–5000, would provide around 5600 additional rebates.

In terms of public relations and brand images, the new challenges in reducing fuel consumption and CO2 emissions have been taken up by carmakers: at major industry exhibitions such as the Frankfurt Motor Show or the Detroit Motor Show, manufacturers great and small are focusing on trying to be "green", displaying concept cars and studies, of which few have reached mass production status, however. The overall atmosphere at the exhibition halls has changed as well: cars are no longer presented in elegant black, but in bright white, and this is no simple attempt to follow trendy electronics producers such as Apple with its white Macbooks and iPods. The aim is to rebrand the entire identity of automobiles as being clean, efficient and environmentally friendly. These are the words that car companies want consumers to associate their products with, which has led several manufacturers to avoid any colors that could be likened to soot, sulfur or other pollutants. Even luxury manufacturers are coming to terms with the idea that it is becoming increasingly difficult to sell cars with above-average fuel consumption and an image as a significant polluter.

In Europe, where the premium car segment is bigger than in the USA, emissions laws have the greatest effect on Daimler, BMW and VW's luxury brand Audi. In 2010, the CO2 figures for the Daimler-brand Mercedes was 172.2 g/km 31.69 miles per US gallon – mpg US), considering only new-car registrations in the EU. In the short term start–stop automatics, efficient transmission and six as well as eight-cylinder direct injection engines should help to reduce emissions. By 2013, the Group plans the new S 500 to come with a plug-in hybrid charging system, the luxury cruiser is supposed to become a three-liter vehicle in the end. This is perhaps the most dramatic example of a general revolution in car manufacturing that will only gain momentum in the years to come.

Daimler's segment competitor BMW, on the other hand, equips all of its models with start–stop automatics as standard: start–stop, a form of hybridization in which the engine switches itself off when at rest, is Hybrid Lite at best. Toyota, by comparison, fits all hybrid cars with two different engines – one gas-driven and the other electric – a system Toyota has also adapted for its luxury brand Lexus. Nevertheless, even this is a far cry from the purely battery-driven roadster produced by the Californian start-up and Toyota-partner Tesla, which is a "zero emission car" under climate change-related laws.

Volkswagen has set itself the goal of reducing CO2 emissions from its new-car fleet in EU countries by 20 percent, from 166 g/km (267.15 g/mi, 32.87 mpg US) as of 2006 to 130 g/km (209.21 g/mi, 41.98 mpg US) in 2015. In 2010, calculations made by external observers put the core brand Volkswagen's figure at 141.0 CO2 g/km (226.9 g/mi, 38.70 mpg US), which will doubtlessly increase if Audi's average 152.9 CO2 (g/km) is included in the Group's fleet average figure, not to mention Porsche.

Meanwhile, the EU's tight limits on emissions have already been loosened by exceptions and regulations for so-called niche products. In Washington, the overall atmosphere regarding vehicle emissions has changed as well. President Obama's new need to seek bipartisan agreement after midterm elections in 2010 has opened the way for blocking actions launched by the Republican Party, which might be seeking to counteract carbon-emissions regulation either by freezing implementation funding for the new regulations or by transferring more of the capacity to make laws on greenhouse emissions back to the Legislative. Nevertheless, in the USA, the long discussions about federal-level emissions regulations and limits are being more and more superseded by concrete emissions standards legislation at state level, where a distinct polarization is emerging between largely Democrat coastal states, led by California, and Republic strongholds in the American heartland.

Even if changes regarding fossil fuel consumption and CO2 emissions might be postponed temporarily they can only be delayed, not reversed. In no small part they are driven by customers themselves. The paradigm of "stronger, bigger, and more comfortable" that was so decisive in automobile research and development for decades has been overtaken by new tenets. The oil price peak during the 2007–8 boom sent a strong signal to car manufacturers in all major markets to take into account fuel efficiency as well as environmental credentials.

At the dawn of the new decade, the CO2 output and fuel consumption, respectively, are the yardsticks which carmakers, especially those based in the "triad" markets, will be forced to let themselves be measured by. Apart from commercial success such as revenue or sales, it is in this field that the two rivals VW and Toyota have sped up the race. Astute visitors to the Frankfurt Motor Show could spot both VW and Toyota representatives curiously eyeing each other's new green models, checking technical displays for CO2 emission data.

Yet this paradigm shift has not become noticeable just in terms of the inflationary production of concept cars to be displayed at ever "greener" motor shows. As the Center of Automotive Management (CAM) based near Cologne, Germany, stated in its yearly innovation study in 2010, R&D "trends show a genuine sea change." Manufacturers worldwide are now conducting three times as much research into alternative energy cars as before.

In its recent study (released in 2011), CAM found that global automobile firms spent a record €50 billion (US$72 billion) in fiscal 2010 for R&D, while 41 billion (US$67.7 billion) had been spent in fiscal 2009. The strongest innovators in the industry are Volkswagen, Daimler, GM and BMW. Powertrain technology is by far the most important field of automotive research and development, comprising 291 innovations and almost half of all evaluated innovations in the 2011 CAM study. Security makes for 15 percent, vehicle concepts for another 14 percent.

Hybrid pioneer Toyota, spending approximately €5.5 billion (US$7.9 billion), and Volkswagen, which spent €5.8 billion (US$8.4 billion), are among the highest spenders in research and development. The continuously high R&D budgets are characteristic of both Toyota and VW, who are clearly ahead of most competitors, few of whom can afford this level of expense.

Contrary to the public focus on green car technology, the number of innovations in conventional powertrain technology is still rising and with 160 they cover the bulk of novelties evaluated by CAM. With 88 counted innovations, hybrid powertrain technology is also gaining ground. Hybrid powertrains are seen as a major bridge technology on the way to the next stage of alternative energy vehicles, while all-electrics still struggle with a series of unresolved technical issues as well as enormous costs. However, only very few manufacturers have sufficient resources to invest in all fields of R&D equally. Innovations in battery-driven cars stagnated – CAM only found 29 novelties.

Nevertheless, CAM deemed the dynamism seen in the electric field in recent years as unique. In its 2007 report, CAM had found almost no notable innovation in electric drivetrains, while in 2009 it booked 30 – even if most of these remain in the conceptual area of R&D rather than series production.

This dynamism appears set to continue in the coming years, and it will undoubtedly have serious effects – not just for the car as a product. The change from fossil fuels to battery-aided or completely battery-driven vehicles has consequences beyond just a serious reorganization for gas station operators.

The gradual replacement of gas-driven engines with other forms of energy will have far-reaching effects on employment, and therefore on one of the core functions of the car industry in its home countries. Currently, hundreds of thousands of workers are employed to assemble ever more complicated engines as well as their components, and this is without taking suppliers into account. Electric motors, however, are far less complicated and will require less manpower in the production phase. Therefore, a major task for car manufacturers will be to secure jobs and to prepare their workforce for new and different tasks once vehicles start rolling off the production lines whose complexity no longer lies in their combustion motors and transmission, but in batteries which may well be supplied by specialist companies.

Currently, it nonetheless remains difficult to say precisely how fast and in which form this next generation of green cars will make for a viable business model, and this is causing considerable uncertainty for the manufacturers. "The market has simply not yet chosen the best low-emissions technology", Toyota-CEO Akio Toyoda puts it – stressing that this is why Toyota is preparing for all options. "When customers do give us their answer, I want the company to be ready", the head of the leader in hybrid technology declared.

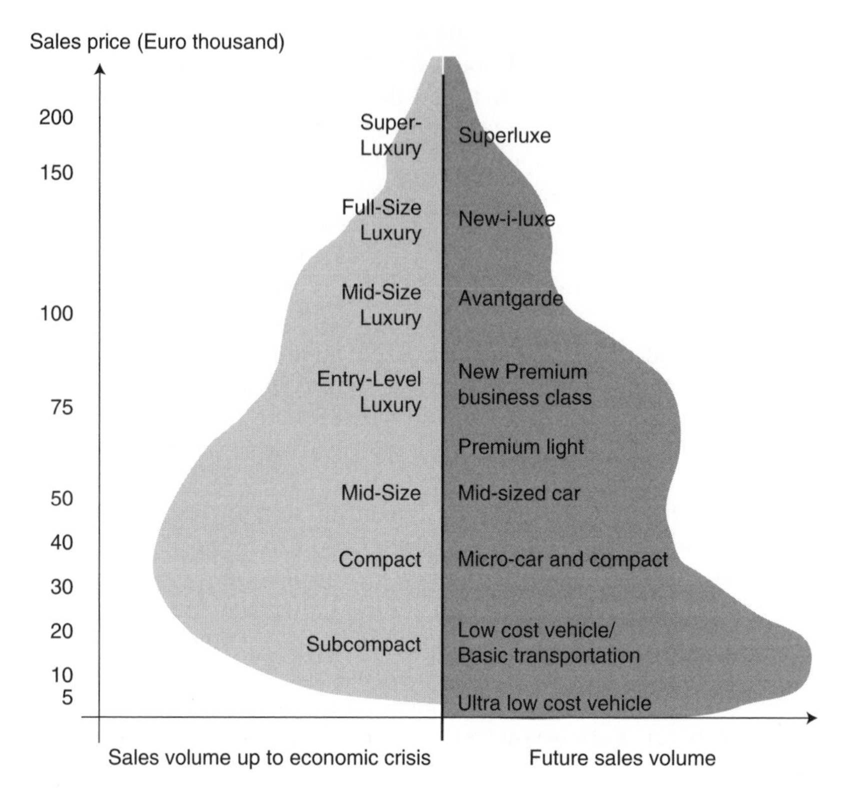

FIGURE 1.2 **Car segments transforming.**

Source: Lecture by Prof. Johann H. Tomforde at the TEAMO Electro-mobility Conference, Stuttgart, Germany, May 5, 2010.

The dilemma is the same for every manufacturer, knowing that when they launch new green technologies, they will lose money for some years before costs fall and volumes rise to let economies of scale make a particular new feature or new technology profitable. The line executive for the 2011 Chevrolet Volt, GM's range-extended electric vehicle, confirmed in a statement in late 2010 that the company loses money on every Volt it sells. Just 10,000 Volts are to be built in 2011, though GM is working to increase that number.

With green cars, in particular, the rate of return is furthermore dependent on factors manufacturers have little control over, such as state subsidies or nationwide infrastructure of energy charging stations.

The profitability calculations for purely electric and plug-in hybrid vehicles presented in this book (see Chapter 2.15 and Chapter 5.5) will make clear that for the customer, buying an alternative-energy vehicle will not always pay off over the car's life cycle. It will most likely be worthwhile in Europe, and in some regions in Asia, mainly due to high gas prices. In the USA, as long as gasoline remains cheap, customers might not want to purchase alternative energy vehicles for at least ten to twenty years, unless they are subsidized (as is the case, there is a federal tax rebate program in place and some Federal States offer EV-incentives on top of that). Looking at the cost calculations over the life cycle of these vehicles, it becomes clear that if these vehicles are be successful on the US market, they will be reliant on high levels of subsidy on the purchase price.

The electric and hybrid strategies of the sector pioneers VW and Toyota will enable these two companies to map the road that other manufacturers will be inclined to follow. Toyota has already done it once with its hybrid systems, which European manufacturers had to see to believe.

TABLE 1.2 **Analysis of segment development (car prices)**

Market segments	Typical example	Comment
Ultra low cost vehicle	Tata Nano Suzuki Lapin	*Here you will also find the famous Japanese K-cars or keijidōsha. They tend to exploit local tax advantages and are really basic cars ("motorbikes with a roof")*
Low cost vehicle/ basic transportation	Dacia Logan Ford Figo	*Mostly targeted on emerging markets*
Micro-cars and compacts	Ford Fiesta Honda Civic VW Polo to VW Golf/Jetta Toyota Corolla Honda Insight	*Wide range for USA, here only sorted by price*
Mid sized car	Ford Mondeo Volkswagen CC Infiniti G37 Chevrolet Malibu Audi A4 and BMW 3 Series Cadillac CTS	*Often referred to as "large family cars"*
Premium light	Lexus GS Lincoln LS Volvo S80 Lincoln MKZ Hybrid	
New Premium Business class	Mercedes E-Class Infiniti M Lexus GS Hybrid BMW 5 Series and Audi A6 Cadillac DTS and STS	
Avantgarde	Maserati Chevrolet Corvette Dodge Viper	
New-i-luxe	Porsche Panamera Mercedes S-Class BMW 7 Series Audi A8 Lexus LS Hybrid Aston Martin	
Superluxe	Bentley Rolls Royce Ferrari Lamborghini	

Accompanied by environmental lawmaking and the unstoppable march of urbanization, another global megatrend will affect the industry: the demand for micro-cars and low-cost models will increase over the coming years. In emerging economies, a customer demographic is emerging which will be able to afford an automobile for the first time, and this will drive the growth in the low-cost segment. In India, for example, models under US$8000 (€5550) dominate the market for personal vehicles; in China, the sweet spot seems to lie between US$10,000 (€6950) and US$11,000 (€7650). The expensive and luxury segments are believed to be growing strongly especially in some Asian regions, while they will increase moderately at best in saturated markets such as Europe or the USA (Figure 1.2, Table 1.2).

The highest margins currently are being made at the Premium levels. However, highest market growth is expected to be at the bottom end of the pyramid, where demand will be strong but margins thin.

In Europe, proof of the growth in the low-cost segment in saturated markets is the popularity of cars like the Dacia Logan, put into the European market at a fighting price of just over €7000 (a little over US$9000). Just two years back, while the major European markets were still being propped up by state-subsidized cash-for-clunkers premiums, the Renault subsidiary Dacia was laughed at as a cheap brand. Now, no one is laughing: Dacia sells the bulk of its cars outside of its home market in Romania. In Europe alone, Dacia sold 214,000 cars in 2009, which represents a 91 percent growth rate since 2008 – and the trend continues unabated. In 2010, Dacia sold 348.000 vehicles, a success which forcefully changed the landscape in the low cost segment in Europe, even after scrappage schemes have been phased out. And there is another competitor who has been gaining more and more ground in this segment – Hyundai.

Hyundai Motor Company

South Korea's Hyundai Motor Company was finally able to shed its reputation as a second-rate brand and now is on the way to become one of the world's most respected and fastest growing auto manufacturers. Churning out some 5.75 million cars in 2010 (including the sister-company Kia Motors) the hidden champion outperformed Ford, increasing its US market share in the aftermath of Detroit's meltdown from 3.7 per cent to 4.4 per cent in just 12 months. In 2011 the combined Hyundai Kia Motors Group eyes record sales of more than 6.3 million cars and, looking from mid-2011, it is well on track to reach this target.

The global expansion of the Korean manufacturer began in the USA in 1986 with the US$5000 car Excel. The cheap car became the top-selling import in the USA in those days, before quality issues caused some fierce headwinds. Conservative planning and luck helped Hyundai to escape the Asian financial crisis of the 1990s relatively unscathed, allowing them at the time to buy Kia Motors.

From the year 2000 onwards, with new management in place, the company rigorously reworked its product line-up and built a reputation for quality and reliability. This has been supported with improved styling and clever marketing around lavish warranty programs. In the midst of the US recession, Hyundai promised to repurchase cars from customers who had lost their jobs, if the car had been sold under a financing or lease agreement. During the troubles surrounding the crisis, this move improved Hyundai's image tremendously: it was seen as a manufacturer that understood the distress of cash-strapped consumers. Hyundai has also developed a green car lineup.

At the other end of the scale, it will be up to luxury carmakers to redefine their models. Beginning in Europe and Japan, most of all, the "new premium" of the coming years will come in the form of "premium light", a class smaller than the current one but which meets the same high customer quality standards. Sheer size might no longer be a reason to buy, and manufacturers seem to have understood this, with Aston Martin building a mini city-car (the Cygnet) on the Toyota's IQ and Audi having launched a luxury mini-car based on the Polo (named A1). Nevertheless, these smaller models must still make as luxurious an impression as their bigger forefathers – and, of course, they must make a profit. Audi even mulls over launching the A1 in India in order to boost its market share on the subcontinent – the trend does not seem to be limited to European markets. All premium manufacturers believe this trend of the "new small premium" will spread out to the emerging markets with some delay but with high sales numbers to be achieved.

Alliances such as the cooperation between Daimler and Renault are further proof of the change in automotive growth markets. Premium manufacturers are forced to trim their offerings in the compact class to make this class profitable, and without alliances, it is difficult to see how this can be done. But high volume manufacturers are also joining forces: Mitsubishi and Peugeot, for example, are also discussing plans for a joint low-cost vehicle for emerging markets. Although a financial link-up between Peugeot/PSA and Mitsubishi appear to be well off the table for now, the automakers look set to continue the ties that they have built up over the past few years – for example a production joint venture in Russia, or the development of electric vehicle powertrain technologies. Toyota is already cooperating with Peugeot's corporate parent PSA on the Aygo micro-car, built in the Czech Republic and sold mainly in Europe. It is in this segment, after all, that Toyota sees its strongest suit.

The Japanese manufacturer has shown that it understands the demands in the micro and compact segments and can comply with them, while making a tidy profit at the same time. The Corolla is a prime example, having become the most-sold car in the world. The Golf, the mainstay model by contender Volkswagen, in comparison to the Corolla is a near-premium product that customers must pay a certain price premium for. Without this extra margin, the Wolfsburg business model has not been functional, let alone profitable. Volkswagen's stake in Suzuki had been intended to help the Group catch up in those lower segments it had neglected before. But more than one-and-a-half years after establishing the relationship there seem to be more German–Japanese cultural difficulties in the cooperation than originally anticipated. So far, results have fallen short of the Group's expectations.

The two rivals VW and Toyota are as contrary to each other as their mainstay models. With a look at their strengths and weaknesses, it becomes clear that the two high volume manufacturers are rivals from two different cultural hemispheres, who are united by success, but by little else.

Toyota is the industry's pioneer in hybrid technology, is skilled at portraying itself as a green manufacturer and wants to fit out its entire fleet with gas-electric engines in the years to come. The German contender, meanwhile, is having difficulty renouncing the purely combustion-engine-based system. Its progress on the hybrid front has been hesitant and is lagging way behind that of its Japanese rival. The first VW hybrid, the Touareg four-wheel-drive, had not been unveiled until spring 2010, and it appears that the Group from Wolfsburg in Northern Germany, like so many in the auto industry, is relying on the prospect that gas-driven engines will, after all, remain the principal form of propulsion for many years to come, and that customers will remain susceptible to falling emissions and fuel consumption for a long time to come. VW's expertise currently lies here, and in other efficiency technology such as weight-reduction.

VW is, for example, ahead in terms of developing ever more efficient diesel and gas-driven engines, being still in line with the overall trend, as earlier CAM-studies showed. In contrast to the manifold high-publicity portrayals on the part of various manufacturers, in the 2009 fiscal year, for example, the majority of the global car industry's innovations relating to powertrain technology were targeted at optimizing petroleum and diesel engines, with about 55 percent of all innovations falling under this heading. Overall, these innovations are no longer about increasing performance, but solely about decreasing fuel consumption.

A look at the global positioning of the two rivals shows that here, too, VW and Toyota could not be any more different. Volkswagen remains strong in its traditional European home markets, with its biggest single market (and clear trump card) being China. Toyota, on the other hand, dominates the Japanese market, but sells the most cars by far in the USA. By comparison, VW has a miniscule market share there and is only just reacting with its new plant in Chattanooga. The USA and China: with their gigantic sales volumes, these two auto markets will decide which company takes the crown in the industry. These two major regions add up to more than 30 million cars in annual sales potential, or around a third of the total worldwide market.

Product philosophy, too, is an area of almost diametrically opposed strategies on the part of the two giants. Toyota is still concentrating on solid, profitable volume in mass markets; very simply put, the Japanese reduce complexity to the minimum, manufacturing reliable and widely successful cars. In contrast to VW, before the crisis Toyota had always had an astonishingly good profit margin on its products. Volkswagen, on the other hand, being a classic representative of the German car industry, sees itself as embodying values such as a high degree of individuality, emotion and customization of each model – for which customers pay a price premium.

The clash between these two rivals, so different and yet so near to each other at the top of this most proud of industries, has electrified not only Volkswagen and Toyota's rivals, but the entire industry, since it was so publicly declared by the Volkswagen boss. The winner of this struggle will present the winning strategy, mapping out which technological trends and product philosophies can lead the car industry into the future, as it moves from fossil fuels to other sources of energy in the coming years.

At this moment of returning profitability – and in the middle of a paradigm shift in terms of green technologies – a key stretch in the long race between VW and Toyota has been entered, and the winner of this duel will most likely emerge sooner than many would expect.

2

VOLKSWAGEN: THE SURPRISE CHALLENGER

It was Martin Winterkorn's crowning moment. On February 3, 2010, the chairman of the Volkswagen Group stepped onto a small stage set up for him in the mirror-walled Paul Hamlyn Hall, the foyer of the venerable Royal Opera House in Covent Garden, London. Sporting a double-breasted suit, the head of Europe's largest carmaker presented his corporation's success story to over 120 investors and analysts gathered there. An additional 400 participants were following his presentation over the Internet.

In front of the grand entrance, Audi's flagship vehicle, a black A8, and a fuel-efficient, silver-blue VW Polo Blue Motion had been prominently arranged on two pedestals. Five white and gray pennants, each printed with the logos of the nine Group brands, from Audi to VW Commercial Vehicles, gave an idea of VW's intricate, multifaceted corporate identity. The location right in the heart of the British capital on this February day provided an impressive backdrop against which Volkswagen would announce its ambition to "become the most successful automotive manufacturer in the world by the year 2018," both commercially and ecologically, as Winterkorn did not tire of mentioning. Volkswagen's strategy calls for nothing less than to conquer the global automotive market with an unprecedented vehicle lineup ranging from a three-liter car and various sports cars and luxury sedans through to 44-ton trucks: it is a plan of attack so ambitious that it has not only stirred up strong rivals such as Toyota, but fascinated the entire automotive industry.

The CEO self-confidently assured his audience that Volkswagen was on an irreversible course toward "profitable and sustainable growth," adding that this clear vision was supported by every single employee and executive in the Group. Moreover, the presentation given by Winterkorn and his Head of Finance, Hans Dieter Pötsch – as is the custom in the car industry – featured the silhouette of a brand-new model out of VW's stables, and remarkably enough the group chose a hybrid concept particularly designed for the US market. The finely styled coupé had already impressed auto experts from around the globe at the Detroit Auto Show in January 2010 and is supposed to challenge Toyota's hybrid flagship, the Prius. The gas-electric VW speedster is based on Volkswagen's bread-and-butter model, the Jetta, an offshoot of the Golf but adjusted for the American market. Compared to the sleek coupé concept produced by the Wolfsburg designers, Toyota's angular and boxy hybrid car appears merely practical. Volkswagen's clear ambition to press ahead with green car technology represents quite a leap for a corporation that has often been sluggish in adapting to change; after all, it had needed nearly three decades to launch the Golf as a successor to the famous VW Beetle, its early best-seller that had become iconic to generations of car drivers all over the world.

2.1 SHAPING A GLOBAL AUTOMOTIVE EMPIRE

In December 2009, Europe's largest car maker took a first step toward realizing German automotive history: it acquired 49.9 percent of the operative business of Porsche AG, the world-famous sports car manufacturer. Entering into agreements with the major shareholder in Porsche, the Porsche Automobil Holding SE, VW spent a total of €3.9 billion (US$5.6 billion) on this gambit. Plans call for completion of the acquisition and subsequent merger with Porsche Holding; the original aim was to create an integrated automotive group in 2011. The step is of historic dimensions as the founder of both Porsche and Volkswagen was the gifted engineer Ferdinand Posche, grandfather of the Chair of the Volkswagen Supervisory Board Ferdinand Piëch. Both companies have been linked together from the start, and Porsche engineers even helped to develop iconic Volkswagen cars like the original Beetle. Nevertheless, they remained legally separate entities; now Piëch is bringing them together.

However, the tight time-frame for completing this merger has been in constant danger of being pushed back. Pending lawsuits have complicated the merger process immensely, with a group of US investment funds seeking about $2 billion in damages for alleged market manipulation by Porsche during its attempt to take the wheel at Volkswagen. In late 2008 hedge fund managers began betting that the price of VW stock would fall. When the financial crisis caused a huge dip in US car sales, driving down the share prices of local manufacturers, they reckoned that it was only a matter of time before German carmakers would suffer the same fate. Consequently, they began selling VW shares in the hope of buying them back later at a lower price – a speculative practice known as short selling. What none of the hedge funds knew, however, was that Porsche had acquired 42.6 percent of VW's ordinary shares with options over another 31.5 percent, bringing its total stake in the company to 74.1 percent. Under German financial regulations, Porsche had long not been under immediate requirement to disclose its ownership of VW stock and options. As a result, hedge fund managers had little idea of the value of the shares they were trading. When news of Porsche's activities broke, the hedge funds were wrong-footed: Porsche's announcement that it had access to almost three-quarters of Volkswagen's ordinary shares sent their price skyrocketing. Within a short time, it shot up from €290 (US$417.60) to €1005 (US$1447.20) per share, making VW temporarily the world's most valuable company. As short sellers – among them a fair number of US hedge funds – scrambled to cover their positions, they incurred huge losses. Consequently, a group of investment funds sued Porsche SE and two of its former top executives, former chief executive Wendelin Wiedeking and former chief financial officer Holger Härter, accusing them of fraud: the funds alleged that they had been victimized when Porsche covertly bought a stake of Volkswagen ordinary shares using swap instruments as part of a planned "sneak attack" to take over Europe's largest automaker – contrary to public statements that it had no plans to do so.

With the outbreak of the financial crisis, however, Porsche's attempt to take over the much larger VW Group failed, and Porsche Holding amassed debts of more than €11 billion (US$15.84 billion), partly due to write-downs on its VW hedges, when it let the air out of their bloated valuation. It became clear that Porsche had dramatically overstretched itself financially, and the economic crisis that was unfolding globally also took its toll – Porsche was in danger of over-indebtedness. Finally, the Emirate of Qatar came to the group's rescue, allowing the sports carmaker to unload its painful derivatives package to Qatar accounts as a matter of last resort.

Thus, now that VW and Porsche plan to combine to build an integrated automotive group, the Arabian Emirate has become VW's third-largest stockholder after the 60 or so Porsche and Piëch family members and the German state of Lower Saxony. For the time being, Qatar owns 17 percent of Volkswagen common stock, while the state of Lower Saxony owns 20 percent; but if everything goes as planned, the Porsche and Piëch families would end up with between 30 and 35 percent of the merged Group. If Porsche Automobil Holding, owned by the families and Qatar, continues to exist as a company for longer than originally intended, its stake in Volkswagen of nearly 51 percent of the voting share would stay untouched.

Meanwhile, Porsche's debt shrank from more than €11 billion (US$15.84 billion) to just €1.8 billion (before the expected inflow of dividends from the investments at Porsche and Volkswagen AG) in the second quarter of 2011. VW had paid €3.9 billion (US$5.616 billion) for its 49.9 percent stake and in April 2011, a capital increase had produced issue proceeds of some €4.9 billion at Porsche Automobil Holding, which was fully utilized to repay liabilities to banks. The controlling shareholders, the Porsche and Piëch families and Qatar, were due to contribute €2.5 billion (US$3.6 billion).

Eventually, in December 2010, a US federal judge ruled that the 39 hedge funds suing Porsche Automobil Holding could not claim the more than US$2 billion (€1.5 billion) in damages they sought from the German company. The US judge also dismissed the damage claims against the two former top executives. The judge referred to a US Supreme Court ruling of June 2010, which said that investors buying shares in foreign companies could not use the American legal system to claim damages for alleged securities fraud. The hedge funds, however, appealed the matter in January 2011.

Meanwhile, German prosecutors have started an investigation into alleged market manipulation and breach of trust by Porsche executives Wendelin Wiedeking and Holger Härter. In early 2011 prosecutors said that those investigations would last until 2012. The former top executives are suspected of having jeopardized the continued existence of Porsche AG by entering dangerous stock-price hedging transactions, thus committing a breach of trust in the course of the 2008 attempt to take over Volkswagen AG. The merger of Volkswagen and Porsche Automobil Holding is predicated on a certain value of liabilities, and the outcome of the investigation could affect the ultimate numbers. In early 2011 the company stated that it thinks there is a 50 percent chance it will complete the merger in time.

On the other hand, there is still another major stumbling block: a potential multibillion euro tax payment. "We're still facing a tax-related hurdle," Winterkorn admitted in October 2010. Porsche acknowledged that potential tax liabilities that could amount to between €1 and 2 billion (US$1.44–2.88 billion) might be triggered by any merger with Volkswagen before 2014. German authorities will have to rule on the issue, which could delay the merger by up to three years or even stop it altogether, according to Porsche and VW.

However, the German carmakers are sticking firmly to their plans to combine the companies, with the completion of the sale of Porsche's core sports car operations to VW as a possible alternative if the initial plan for a merger between Porsche Holding and Volkswagen doesn't materialize. "The integrated automotive group will happen," Winterkorn emphasized in October 2010. Porsche and Volkswagen have granted each other put and call options to transfer the remaining 50.1 percent stake in Porsche's sports car unit to VW between November 15, 2012 and January 31, 2015, in case the merger cannot be completed by the end of 2011 as anticipated.

For Porsche, a significant future benefit of any merger or combination will be better access to Group technologies – an important key to expanding its global sales. In the medium term, Porsche is targeting annual sales of 150,000 units, after only 82,000 in the 2009/10 fiscal year. With effect from January

1, 2011 Porsche AG has aligned its financial year – which had previously run from August 1 to July 31 – with the calendar year; in the short financial year from August 1 to December 31, 2010, Porsche AG significantly increased its sales. Compared with the same period last year – namely August to December 2009 – global sales grew 57 per cent to 40,446 vehicles. As a next step, a fifth production series – alongside the 911, Boxster/Cayman, Cayenne, and Panamera models – is intended to fill the segment below the Cayenne. The small SUV, currently named Cajun, is based on the architecture of the Audi Q5. In addition, VW's managers see market potential for a small sports model with a charged four-cylinder engine built in a cost-saving manner on the basis of the Group's pre-defined range of parts and modules. This product would be positioned below the Boxster, the previous entry-level model, with a list price starting at around €40,000 (US$57600).

Conversely, Volkswagen also wants to benefit from Porsche. The idea is for some of the brand's charm – it has an almost cult-like following – to rub off on the Group's high-volume models, as Lamborghini has already done for Audi: dealership salesmen like to entice customers by pointing out that the ten-cylinder engine used in Audi models is a Lamborghini unit. Jointly developed high-performance brakes could be marketed as Porsche brakes, and further synergies like this would boost the image of other Group brands as well. In the future, all of the Group's sports cars will contain modules developed by Porsche, and the Porsche Development Center in Weissach, Germany, will define the future course of sports car building and develop Group-wide basic technology to this end. In light of its own success with modular longitudinal and transverse toolkits, VW has assigned Porsche the task of developing a high-end modular toolkit for sports cars as well as VW's luxury brands, an assignment that certainly strengthens the formerly independent carmaker's Group role as it is gradually folded into VW's operations. VW estimates the total value of synergies between the two carmakers at €3 billion (US$4.32 billion), arising at about €700 million (US$1 billion) per year mainly from joint purchasing, administration, financial services, and the all-important technology sharing.

To finance the integration of Porsche while preserving the Group's investment rating, Volkswagen's finance chief Pötsch put together the largest capital increase in German economic history: in March 2010, VW issued just over 65 million new shares of preferred stock, bringing in net proceeds of €4.1 billion (US$5.9 billion). At the Group's annual stockholders' meeting in 2010, Pötsch also approved the sale of up to €5 billion (US$7.2 billion) in convertible bonds, so that the Group gains flexibility with an additional financial instrument. That was necessary because corporate rules do not allow VW to issue more new shares than already approved: following this, he was unable to issue any further preferential shares for six months.

As a financial basis for the integration of Porsche, the Group's automotive business had liquid assets of over €20 billion (US$27.5 billion) as of May 2011. In total, the entire deal – including the acquisition of Porsche's automobile trading business from the Porsche and Piëch families – will cost Volkswagen at least €16 billion (US$23 billion).

In addition to the Porsche deal, VW managers want to move forward with the strategic alliance among the Group's commercial truck holdings, a project which had been stagnant for years. This effort will be led by former production chief and management board member Jochem Heizmann, who has managed the Group Commercial Vehicles unit since autumn 2010. Volkswagen, which owns nearly 71 percent of the Swedish truck maker Scania's votes, also has a voting stake of nearly 30 percent in German-based truck maker MAN, which it increased to 30.47 percent in early 2011, requiring VW to make a mandatory bid for the entire company – a step that facilitates the merger between MAN and Scania.

Volkswagen has made no bones about its desire to reap synergies from having the truck makers join forces. Chief VW supervisor Ferdinand Piëch is on record as wanting Scania to enter into greater technological cooperation with its German competitor MAN, and despite legal impediments, VW has been taking the lead to see this project through. In May 2011 Volkswagen stated that the company aims to raise its stake in MAN from 35 to 40 percent as a first step to gaining merger control clearance, but it is not aiming for a majority stake right away. The two companies had for some time been mulling over various possible industrial projects that would generate savings in research and development, manufacturing and sourcing. However, a full realization of potential synergies requires a closer cooperation by combining the two companies, and so the crux of the matter is therefore permission from the relevant competition commissions.

The two companies will almost certainly have to join forces in some way or other, if they are to survive in global competition with market leader Daimler, compared to whom MAN and Scania are relatively small players. Combined, MAN and Scania are already strong in Europe: together they had 30 percent of the European heavy-truck market in 2010 (according to the European Automobile Manufacturers' Association), whilst Volvo and Daimler each had 21 percent. Furthermore, MAN has stated that it too understands the logic of closer cooperation with Scania and VW. Experts see cost reductions of about €1 billion for an integrated company fully sharing research and development activities; VW sees savings of at least €200 million by combining its already monumental purchasing power that of MAN. The two truckmakers face difficulties, however, as EU cartel authorities in early 2011 launched a series of raids on major companies in the trucks segment in Europe, including MAN and Scania. EU cartel investigations typically take years and the uncertain outcome might also complicate the possible tie-up between the two.

Another impediment was the opening of a corruption probe into MAN's former subsidiary Ferrostaal. The International Petroleum Investment Company Ipic, as the Abu Dhabi investment fund is known, bought 70 percent of Ferrostaal from MAN for about US$650 million in 2009. Since then, however, Ferrostaal has become the target of a corruption investigation in Germany over contracts in several countries stretching back decades. The ongoing bribery investigation relates to the time when Ferrostaal was still part of MAN. MAN is now locked in a dispute with Abu Dhabi's International Petroleum Investment over the costs of the bribery scandal, and Ipic is trying to block a provision of the Ferrostaal deal requiring it to buy MAN's remaining 30-percent stake.

Despite those drawbacks, Winterkorn and his mentor Piëch seem to have nearly reached their goal after three decades of joint effort – an automotive empire that spans all car classes and segments from micro-cars to trucks. "He comes up with the innovations, and I ensure that they happen," Winterkorn says about his relationship to Piëch. As long as these influential VW personalities do not encounter any serious mishaps, they could very well attain their ambitious sales target over the coming years: at any rate, the top spot in the automotive industry seems nearer than ever.

German company law

In German corporations, the functions of the board of directors are split between a management board (*Vorstand*) and a supervisory board (*Aufsichtsrat*). VW CEO Martin Winterkorn is Head of the management board, while Ferdinand Piëch, the VW patriarch, is Chairman of the supervisory board.

In large companies subject to laws regarding employee participation, such as Volkswagen, one-half of the supervisory board is elected by shareholders and the other half by trade unions and employee organizations. The supervisory board then elects the management board, which is charged with running the company. In theory, the members of the supervisory board are responsible for holding the management board to account.

The power invested in the chairperson of the supervisory board, Ferdinand Piëch, is immense. In larger listed companies, where 50 percent of the board is composed of employee representatives, it is the chairperson who holds the casting vote in the event of an even split. The supervisory board, however, is obliged to take account of the provisions of the German Corporate Governance Code, but may choose to deviate from these in exceptional circumstances.

Winterkorn and Piëch's starting position for a tie-up is more promising, with the 2010 fiscal year proving to be the best year in the history of the Volkswagen group. The Germans sold a record 7.14 million units, a 13.5 percent increase compared to the prior year, while for 2011 the Group targets the 8 million sales. In 2010, VW had record net profits of €7.2 billion (US$9.9 billion), a jump of over 700 percent compared to the previous year. Net profit got a boost from VW's strong presence in China and other emerging markets, financial investments and the value of options related to its holding in luxury sports car maker Porsche, "We believe that 2011 turnover and operating profit for the company will exceed the preceding year's tallies,"Winterkorn announced. In 2010, VW operating profit soared to €7.1 billion (US$9.8 billion), more than three times the figure from the prior year (€1.9 billion, or US$2.6 billion), on sales up 21 percent to €126.9 billion (US$174.5 billion).

Winterkorn and Piëch simply will not accept coming second: the two want to reach the very top of the automotive industry. "VW is starting from pole position. We are driving this race with the requisite aggressiveness, and we will take great care to stay right on track," the CEO promised stockholders in 2010. Making a compromise is not an option: "We have our sights firmly set on our long-term objective to take the Group right to the top."

2.2 STRATEGY FOR REACHING THE TOP

Leading in automotive sales alone, however, would not safeguard a top spot in the industry – Toyota's quality crisis has shown the German car maker that it cannot focus on size alone. Rather, its assault on the position of global market leader is based on four objectives that Volkswagen wants to achieve by 2018 at the latest: annual sales should grow to over 10 million vehicles; the operating profit margin before taxes must top 8 percent; the corporation should assume a leadership role in customer satisfaction and quality; and Volkswagen wants to be perceived as the best employer in the entire industry.

The VW chief sees the main factors in his company's success as a combination of Italian styling, delivered by colleague Walter de'Silva as Head of Design, and high engineering quality, which Winterkorn himself guarantees and measures by German standards of craftsmanship. The new Polo – which launched in 2009 and was jointly developed by Piëch and Winterkorn right from the start – has been praised by industry experts and awarded such laurels as the "Golden Steering Wheel" and "World Car of the Year."

In his quest for perfection for the entire Group, Volkswagen's top manager Winterkorn has identified several main lines of attack designed to guide VW to the top and making it the "leading car manufacturer, both in economic and environmental terms" as he puts it:

- focusing on growth markets
- modularizing models
- disciplined employment of capital
- increasing operating profit
- leveraging potential synergies between brands.

Focusing on growth markets

In order to boost sales, Volkswagen wants to penetrate saturated markets better and conquer new growth regions.

For the expansion strategy to pay off, the manufacturing network must grow with sales – and it must do so globally. Volkswagen cannot cover its costs and deliver high-volume models to emerging countries by mainly producing in Wolfsburg and Western Europe. The car maker must follow market demand. At the press briefing on the 2010 annual results in Wolfsburg, in March 2010, VW Labor Director Horst Neumann noted that at least half of the Group's approximately 369,000 employees work outside of Germany, and in 2018 that proportion is planned to reach 60 percent.

Above all, this change is being driven by China, which has now become Volkswagen's most important sales region. In China, Volkswagen has stakes in two joint ventures with government-owned companies: Shanghai Volkswagen Automotive in the country's southern region since 1984, and FAW Volkswagen in the north of the country since 1990. In 2010, Volkswagen employed a total of about 41,000 people in China and operated four production plants for vehicle assembly and four more for components. Over 1600 employees working in technical development make the People's Republic the Group's second most significant research and development center after Europe. Furthermore, the automotive market in China is booming: in 2010, VW's deliveries in China grew 37.4 percent to 1.92 million units, thereby setting a new record. The Group intends to increase its annual production capacity in China to about three million vehicles by no later than 2014, and four new factories are to be built. Between 2011 and 2015, the two Chinese joint ventures will invest a total of €10.6 billion (US$15.264 billion) in their automotive business, according to VW. The importance of the Group's relationship with China was reflected in an article by the VW chief that appeared in China's English-language newspaper *The China Daily* during the 2010 Beijing Motor Show under the fitting title "VW Group and China: A friendship of giants."

Currently, Volkswagen's corporate empire spans more than 60 production sites – not including the two Porsche sites and additional factories currently located in China. Each workday, the assembly lines crank out over 30,000 vehicles worldwide. The company has benefited from the fact that its managers pressed ahead with internationalizing its manufacturing base early on. In the 1950s and 1960s VW began to produce vehicles in South America – in Brazil in 1953 and Mexico in 1964 – before China was added to this list in the 1980s. The acquisitions of SEAT and Skoda integrated manufacturing facilities in Spain and the Czech Republic into the production network, and now Volkswagen is upgrading its plants in Russia and India, which opened in 2007 and 2009 respectively, into full-fledged production facilities as soon as possible. The company's approach to growth markets has been consistent: initially, workers assemble kits from Group brands VW, Skoda, and in some cases Audi, then the Germans launch full production in these markets.

Modularizing models

Volkswagen is setting great store by modular toolkits, which, used as the basic architecture of a wide variety of car models, are slated to generate enormous corporate-wide savings in capital expenditure, development, and production costs. If these plans come to fruition, each of the automobile brands under the VW umbrella would benefit from previously unimaginable economies of scale. The standardization of key components such as axles, transmissions, and climate control systems is intended to accelerate production – a weakness at VW with its distinctly average efficiency – while shortening development times for new models and simultaneously reducing costs.

Audi's success story shows the potential power of technological innovation: Piëch, first as Head of engineering and then as Audi CEO, steered the car maker toward premium quality from the early nineties on. The plan was for the VW subsidiary Audi, based in Ingolstadt, Germany, to gradually position more and more high-end vehicles against BMW, for VW itself to measure itself against Mercedes as a premium brand, and for Skoda to rival Volvo. The first generation of the Audi 100 had already managed to penetrate the upper middle class segment by 1968 – even before Piëch's time. In early 1980, Audi presented the first all-wheel drive Audi Quattro at the Geneva Automobile Show. Groundbreaking technical innovations – such as the first direct injection diesel engine (TDI), which made its debut in the Audi 100 in 1989 – made the brand appealing to customers and paid off for the Group by being used across brands.

Under the auspices of Piëch, Audi systematically assumed the role of the pioneer within the Group. As early as 1994, Audi arrived in the luxury class, with its new A8 sporting an aluminum body. Since 1971 the brand has taken

"Vorsprung durch Technik" ("Advancement through Technology") as its fitting slogan and has simply not tired of introducing technological improvements into almost every new design. The new 2010 edition of a 2.5 liter, V6 TDI, impressed with the lowest fuel consumption in its class: in the new European Driving Cycle (NEDC), the car consumes six liters of fuel per 100 kilometers (39.2 miles per gallon) and is "by far the most fuel-efficient luxury car in the world," according to the German automotive press. In Wolfsburg, Piëch has long made Audi's premium aspirations a role model by offering more and more models with increasing quality, all the while standardizing components across production series. However, he constantly foments competition between the individual brands under the VW umbrella.

Disciplined employment of capital

Despite the plan to increase sales volume, VW must not let costs get out of control. After Porsche's failed attempt to take over Volkswagen, the Group now needs to absorb billions of Euros in costs for integrating the sports car maker and Porsche's huge automobile trading business, the Salzburger Porsche-Vertriebsholding. Excluding Porsche, VW wants its automotive business to reach a Return on Investment (RoI) after taxes of over 16 percent.

Just a few years back, this seemed like a very tall order. In 2008, RoI was at just 10.9 percent; one year later, in the midst of the global economic crisis RoI in the automotive area hit an alarming 3.8 percent – too low to recoup its capital costs. In the Group's automotive business, VW estimated invested assets for 2009 to average €43.1 billion (US$62 billion). At an assumed average cost of capital of 6.9 percent, VW incurred nearly €3 billion (US$4.3 billion) in capital costs, which more than offset the operating profit after taxes of €1.6 billion (US$2.3 billion), resulting in a €1.3 billion (US$1.9 billion) loss in the automotive business.

In 2010, however, RoI more than tripled to reach 13.5 percent, putting the 16-percent mark suddenly well within Volkswagen's grasp. In the Financial Services business meanwhile, Winterkorn is targeting a Return on Equity (RoE) of 20 percent before taxes. In 2010 RoE was 12.9 percent, bouncing back from 7.9 percent in the crisis year 2009.

Improving operating profit

Piëch and Winterkorn, two engineers keen on attaining technical perfection, drive their developers to deliver high performance. Winterkorn, for example, demands that the feel of a leather steering wheel must be identical to that of the leather-trimmed gearshift grip, even though the two parts come from different suppliers. Acceptance test drives, during which the two top managers might test the new vehicles on ice or in the desert before releasing them for sale, are dreaded internally. An uncompromising passion for automobiles is a core value in Volkswagen's headquarters, as expressed in the VW brand's advertising slogan, which is reiterated globally in German words: "Das Auto" (transl. "The Car"). Without this high standard, no cars could be built in Germany, Winterkorn emphasizes.

To offset local production disadvantages and the comparatively high costs of preserving its key assembly plants in Western Europe, Volkswagen has been pursuing a particular approach to optimizing processes for several years now, which is referred to as the "Volkswagen Weg" (transl. "The Volkswagen Way"). Continuous Improvement Process, or CIP, workshops, based on Toyota's role model, are intended to make the Group's factories more competitive. The goal is to increase productivity by 10 percent per year. Progress has been made, such as with the new Golf, where the key to higher productivity, and thus to a higher margin, was a decisive change in strategy that yielded unprecedented savings: existing technologies and tools from previous models were re-used intensively in the newer model. During previous changeovers, the entire body-in-white was replaced, regardless of whether or not the facilities and tooling used had already been amortized. Engineers now make running improvements to components such as axles without redesigning them from scratch. Hence the current Golf model, for example, which – due to the inadequate profitability of the previous model – was launched early on the market in October 2008, is more an extensive vehicle facelift than a completely new car.

While the handsome profits in recent years often were offset by high indirect costs, tactical sales campaigns or expensive bail-out measures for ailing suppliers, VW surprised the entire industry with its good results in 2010. Although sales only went up by 11 percent for the core VW brand, sales revenue went up by 23 percent and operating profit almost quadrupled, albeit from a very low starting point: in 2009 operating profit was a measly 0.9 percent, in 2010 it had jumped to 2.7 percent. While 2.7 percent would not be an operating profit worth mentioning for some other manufacturers, it certainly showed detractors of the Group that there is potential in its operations – and that it is able access this potential. Other Group companies surprised observers too, with Audi more than doubling its already far higher operating profit between 2009 and 2010. Further more, at the time of writing in 2011, this seems more like a consistant trend upwards than a one-off blip: operating profit across all Volkswagen brands was running at 7.8 in the first quarter of 2011, compared to 7.1 for 2010 as a whole and 1.8 just two years earlier in 2009. In terms of its historically poor profits, the Volkswagen Group – and especially the core brand VW – is making progress in leaps and bounds, and is currently in something of a virtuous circle: as its sales volume climbs, so does its power in purchasing. Meanwhile, continuing high demand in China is seeing factories running at 130 percent capacity and keeps sales prices stable. VW is also reaping the benefits of its relatively new range of high-end models with plenty of amenities and extras, which bring in a high margin – and even on these models, as with the new Golf (which may even already appear in 2012), Volkswagen has been shortening its model cycles and developing its vehicles in a far more cost effective manner.

Further to this, with all this growth, expenditures for complex special projects such as the "Autostadt" (transl. "auto city") tourist attraction, the "Autouni" – a management academy – and the German national soccer league team VfL Wolfsburg, are more negligible than in previous years. Such activities are not being terminated – being as they are characteristic for Volkswagen as a company and as a brand – but the long shadow they so often cast on the balance sheets is receding. The same is true of the high overhead costs it has, especially in view of the pace at which has been opening new factories: in bad times, investors would complain about this at annual meetings; in times like these, this kind of outlay can be overlooked, especially since it contributes to the company's ability to keep growing.

Another characteristic of the Volkswagen organization is its vast depth of in-house production, i.e. the extent to which individual car parts are produced by VW-owned facilities. VW has in-house component facilities for axles, transmissions, and car seats, for example, keeping more than 15,000 employees busy in VW-owned plants; about the same number of employees produce other vehicle components such as bumpers. In-house supply of parts can come with cost disadvantages – especially when external suppliers such as world market leader Bosch can manufacture standard parts far cheaper by aggregating production volume across several original equipment manufacturers (OEM) – and in previous years VW's production depth was often cited as a major reason for its poor productivity. Yet this kind of depth in value creation can turn out to be a strategic advantage when it comes to technically highly specialized components such as the innovative DSG dual-clutch transmission, which is a flagship innovation in VW's empire and is of course supplied by VW-owned component plants. Above all, in the periods of expansion such as the car industry is currently undergoing in 2011, innovations such as the DSG are making VW cars attractive, and so this kind of in-house components expenditure is not only balanced out by high sales and profitability elsewhere in the company, but can actually be said to be a motivating factor behind them. Essentially, as long as Volkswagen's sales are as high as they currently are, it can afford its complicated structures.

Nevertheless, VW's particular mix of high standards and expensive cost structures remains a drawback for the company, particularly in the small car segment, which in recent years has grown to at least 40 percent of new car registrations in Western Europe and promises further gains worldwide in coming years. Volkswagen does not appear to be in a position to build cars priced below €5000 (US$7200) within its corporate and cost structures – a shortcoming which is slowing the Group's progress in emerging countries such as India. To address the problem, the Group established a joint venture with Japanese producer Suzuki in December 2009, but it is still unclear what will come of the cooperation (see Section 2.13).

Potential synergies

Volkswagen's management wants to pick the most promising solutions generated by the internal competition among its various brands and roll them out Group-wide. It also aims to benefit from joint purchasing, combined production, and synergies in sales and marketing. The Group can particularly benefit from its size when it comes to expensive investment projects such as green car technology. VW can aggregate basic technology across a number of high-volume Group-owned automakers while other companies – such as Audi's German competitors BMW and Mercedes – must turn to external alliances in order to reduce costs significantly, as the cooperation between Daimler and the French high-volume producer Renault illustrates.

Meanwhile, Volkswagen can draw upon a broad base of experience acquired over many years of successfully managing its various brands. This began back in 1965, when the acquisition of the multibrand corporate group Auto-Union from Daimler-Benz put a second brand into the hands of VW managers in the form of Audi. In the mid-1980s, SEAT was added in Spain, and after the fall of the Iron Curtain, Skoda in the Czech Republic followed in 1991. In the late 1990s, under the auspices of Ferdinand Porsche's grand-son Piëch as Chairman of the Board, the VW Group purchased the luxury brands Bentley of Great Britain, Bugatti of France, and Lamborghini of Italy. In early 2008, the Group acquired a majority stake in the Swedish truck maker Scania as brand number nine (if the Volkswagen Commercial Vehicles line is considered as a separate entity). Volkswagen seeks to benefit from a unique product lineup that ought to number ten brands when sports car maker Porsche slips under the covers of VW. With Audi, VW, Skoda, and SEAT, the Group has a fair number of passenger car labels with high-volume sales (Table 2.1).

Winterkorn first formulated these elements of his "Strategy 2018" publicly for investors and analysts in London on February 3, 2010; and little more than a year later, it would seem that progress has been made on all of the points mentioned. The Group can point to its focus on China as a growth market, to continued modularization and discipline in employment of capital, to vastly increased operating profits and to progress in synergies. Winterkorn also

TABLE 2.1 **Brands and acquisitions of Volkswagen AG**

Year	Majority Stakes/ Acquisitions	Comments
1937–38	**Volkswagen**	Founding of the company Gesellschaft zur Vorbereitung des Deutschen Volkswagens mbH
1939		Company is renamed Volkswagenwerk GmbH; Franz Xaver Reimspieß designs Volkswagen's brand logo that is still in use today: The letters V and W enclosed by a circle.
1965	**Audi**	A brand group is started with the acquisition of Auto Union GmbH and NSU Motorenwerke AG.
1969		Auto Union and NSU merge to form Audi NSU Auto Union AG.
1985		Audi NSU Auto Union AG is renamed Audi AG.
1986	**SEAT**	Majority stake acquired in SEAT. After Fiat pulls out of the Spanish company, Volkswagen fills the gap. Over the course of subsequent restructuring, SEAT is integrated in the Volkswagen Group as a third independent brand.
1990–91	**Skoda**	Acquisition of Skoda, automobilova a.s. after fall of the Iron Curtain. Skoda is intended to guarantee VW good access to automotive markets in Central and Eastern Europe. The Czech-based automotive company becomes the fourth independent brand in the growing Volkswagen Group.
1995	**Volkswagen Commercial Vehicles**	Creation of a label for light trucks.
1998	**Bentley**	After a bidding war with BMW, VW buys the British company Rolls Royce from the Vickers Corporation. Although Volkswagen does not acquire rights to the Rolls Royce name, it does acquire the automotive plant in Crewe and the name Bentley. With Bentley, VW begins to extend its product lineup into the luxury segment.
1998	**Bugatti**	Volkswagen Group France acquires the French vehicle brand based in Molsheim, Alsace. In 2000, Volkswagen founds Bugatti Automobiles S.A.S. This is the eighth brand in the corporate group.
1998	**Lamborghini**	Audi AG takes over the Italian sports car manufacturer.
2006	**MAN**	VW acquires a strategic stake of 15.06 percent in truck maker MAN AG. This secures the interests of the Volkswagen Commercial Vehicles brand in the strategic partnership initiated between MAN and Scania.
2008	**Scania**	VW acquires a majority stake in Scania and enters the global truck business (original equity stake of 37.73 percent, voting rights share of 68.6 percent, now at more than 45 and 70 respectively).

Continued on next page

TABLE 2.1 (Continued)

Year	Majority Stakes/ Acquisitions	Comments
2009–10	**Porsche**	Acquisition of 49.9 percent of stock in Dr.-Ing. h.c. F. Porsche AG represents a milestone in creating an integrated car group with Porsche.
2009–10	**Suzuki**	VW acquires a 19.9 percent stake in Suzuki Motor Corporation for about €1.7 billion (US$2.4 billion) with the goal of a long-term strategic partnership. Suzuki wants to reinvest half of the purchase price in VW stock.

Source: Navigator 2010 – Figures, Data, Facts, Volkswagen AG, compiled by authors.

spoke in London about "exploiting upward potential" – and many see this as code for what some Volkswagen executives see as the Group's biggest piece of unfinished business: America.

2.3 SHOWDOWN IN THE USA

2011 will go down in Volkswagen annals as the crunch year for its US operations. It will either be the year in which Volkswagen finally turned around its fortunes in this crucial market, or the year in which it massively over-invested in a hubristic enterprise doomed to failure. Only time will tell. Since its founding 1955, Volkswagen's US experience has been double-edged, swinging wildly between victory and defeat. Indeed, Volkswagen of America later proved to be a lucrative source of profits. Under the leadership of former CEO Carl Hahn – and thanks to the Beetle – Volkswagen attained a 7 percent market share in 1970, with a record of 570,000 new car registrations. In 1988, however, the only American VW factory (in New Stanton, Pennsylvania) had to close its doors after ten years of production. Five years later, Volkswagen reached rock bottom in the USA, selling just 50,000 cars. Since that time, the sales curve for the Group has been similar to a roller-coaster ride. The quality of VW's image, customer loyalty ratings, and sales figures have disappointed year after year. In 2009, Group sales in the USA did not even reach 300,000 units, making for a miserable market share of just 2.9 percent. Since then, VW has gained ground: Volkswagen of America closed 2010 with the best overall year sales since 2003, the Group brands delivered 360,300 units during the 12-month period, corresponding to growth of 20.9 percent. The year 2011 is shaping up to be a good one, but it will take a few years for analysts to be sure that the roller-coaster ride is over.

The automaker's own long-term projections are positive, even though the US division has failed to return any profits to Wolfsburg since 2002. Volkswagen has made the most of a difficult situation and for many years focused on the Chinese market instead; the company expects growth in China to accelerate it to the position of the world's leading carmaker. Nevertheless, VW has no choice: in view of the lasting importance of the USA, it must make big gains in the second most important car market in the world if it is to catch up to rival Toyota. This importance explains the lasting fascination in Wolfsburg with the possibilities of the American market, and the decision to spend so much money on a new factory – and an almost complete relaunch of the company to accompany it – in such a historically difficult market. The new beginning is most visible at Chattanooga, yet in the same year as construction started on this gleaming new production facility (2008), Volkswagen also took another telling step, moving the North American company headquarters in 2008 from Auburn Hills near Detroit to the Washington suburb of Herndon, Virginia. With this move, the company abandoned its local industry counterparts – the domestic US car makers in Detroit – and moved closer to the political decision-makers. At the same time, the move was symbolic: the company was moving closer to its geographical customer base, which is concentrated primarily on the east and west coasts.

Symbolism and customer dialogue are important parts of Volkswagen's fresh start in the USA. The goal now is to "show the USA that we are an American company," as Winterkorn says, obliquely referring to Volkswagen's recent ignominious history of trying to force-feed the American market with essentially European models – models made expensive by things that Europeans might be willing to pay a lot of money for, but that US car buyers see as pointless. For years, the Group's executives overloaded their cars with technical innovations that Americans just didn't want to buy; and when American motorists didn't buy the cars, the Group assumed that they simply hadn't understood them. Audi's reaction to as series of "runaway car" issues in the USA in the early 80s, for example, spoke volumes about the German automotive Group's attitudes towards the US market: a series of dismissive press releases and comments both on and off the record made it clear that, as far as the automaker at Ingolstadt was concerned, the problem was not the cars, but that Americans didn't know how to drive them. Now, VW is listening to American motorists and building them the large, uncluttered cars that they want to drive, but with a European – i.e. fuel efficient – twist; and this may be just the right time to be offering this kind of vehicle. Given recent changes in the tastes and demands of American consumers, the VW chief sees his company now well positioned for success in this market because of its strength in "large, but fuel-efficient cars."

The Volkswagen Group has set a sales target of one million cars in the US market for 2018. Of that figure, VW estimates that Audi will sell 200,000 units. The German car maker is relying on a four-pillar strategy: developing products specifically for the US market; re-positioning its brand; changing almost half of its dealership network – a process that has been ongoing since 2008 and should be finished by 2012 – and finally, pushing its own production in the USA or at least in the NAFTA area. The site in Chattanooga is to churn out 150,000 units per year.

Stefan Jacoby, President and CEO of Volkswagen of America since 2007, had been selected to lead this offensive, but then irritated the carmaker in summer 2010 with his decision to join the competition. Jacoby is now CEO of the Swedish automaker Volvo, which Geely Holding Group, a Chinese company, purchased from Ford. In October 2010, Jonathan Browning took over as VW's new helmsman for its US operations. Browning, a GM veteran and former vice president of global sales, service, and marketing for the Detroit rival, initially joined the Volkswagen team in June 2010 to head the national sales effort for all nine brands under the Volkswagen umbrella. Taking up his office at the VW headquarters in Virginia, Browning acknowledged that he faces a challenge, since the company has not managed to surpass its own 30-year old benchmark: in 1970, VW sold almost 600,000 vehicles in the USA.

For his part, Browning declared that he was focused on the successful launch of three new vehicles geared toward the US market: first, there is the all-new Jetta, built in Puebla, Mexico. The car showcases the German manufacturer's will to win over American customers with a decidedly American sense of styling and prices competitive with core competitors such as Toyota. The growing popularity of the new model confirms the strategy: since its launch, the Jetta has greatly contributed to increasing VW's sales in the USA. Second, there is the new Passat which debuted at the 2011 Detroit Motor Show, and is positioned to challenge Toyota's successful family car, the Camry: it is priced under its Japanese rival, retailing for just US$19,995, or under €15,000 in Volkswagen's home currency. The idea of selling a mid-class sedan like the Passat at such a low price would have been anathema to Wolfsburg executives a few years back, but they have finally understood that Americans just will not spend more and have adapted the vehicle accordingly and are now well-positioned; annual sales in the mid-class sedan segment amount to a promising two million cars in the USA. Third, VW is to launch a next-generation Beetle slated to follow the sedan in 2012. According to Browning, the company expects the United States to be the highest-volume market for the Beetle – an iconic model still reminiscent of VW's decades old sales momentum story. With higher gasoline prices and the consequent trend toward smaller motors is coming a revived interest in German cars generally in America, and as long as Volkswagen plays by the rules of US market and keeps the prices down, it might even stand to profit from the slightly more quirky, cult side to its image in America.

Nevertheless, the American market has largely been propped up by incentives and sales support for over ten years now – especially in 2010 – and there is no reason to expect the market situation to change in the foreseeable future. On the contrary, applying pressure on Toyota in the USA led Toyota to increase its relatively conservative rebate levels. Although the rebates did level off again to the pre-crisis levels of 2008 by the end of 2010, this business environment leaves little opportunity for winning over customers with innovative products: it's a difficult starting situation for Volkswagen.

Another question that remains open is that of parts supply. To actually reap the advantages of the dollar market with the new factory, an estimated 85 percent of car parts must be sourced from the region as quickly as possible. This is an enormous challenge, since it makes a big difference to VW's suppliers whether they are supplying parts for a million VW vehicles in Europe or for a mere 150,000 in the USA. Localization can succeed nevertheless, but it will take a long time and will place extreme demands on resources that Volkswagen must provide from its European home base. Even if this localization effort succeeds, the Group will have to continue to reposition its brands in the USA. Another reason its orientation in Europe toward high quality and technological finesse has not been successful in the USA is because the Group's vehicles are not seen often enough on US roads and highways. If it really wants to hit the million mark, Volkswagen must continue to create a new and credible identity in the USA, escaping its image as the hippie-inspired manufacturer of low-horsepower cult-like cars and growing into the role of a carmaker who confidently caters to the mainstream.

However, Americans take a very different view of what practical, everyday cars with low maintenance costs look like. Former US Chief Jacoby had to concede that "Volkswagen, like all European brands in the USA, is considered expensive in customer service." From his perspective, Volkswagen had positioned itself for too long in the "niche for exotic cars." Volkswagen currently has roughly 600 dealerships in the USA, more than half of which are exclusive, and has properly invested in the VW showroom concept. Since 2003, dealers have spent more than US$3.5 billion. The idea is for the money to finally flow back to them now. Volkswagen is making its cars and brand attractive by improving its image and lowering prices. For example, all models include free maintenance and service for the first three years after purchase. VW uses tactical advertising to boost its currently low brand strength – notably in professional sports, which have high emotional appeal in the USA – and improve brand recognition among the US middle class. This includes sponsorship, e.g.

of soccer clubs such as DC United, that enjoy especially high popularity among Hispanic Americans. In early 2011, one of the Super Bowl commercials everyone talked about most was the Volkswagen ad in which a pint-sized Darth Vader attempts to use the Force – and only finds success with a new VW car.

In VW's ad, a young boy is marching around his house in a Darth Vader costume. He tries to use the Force to get his dog to rise, open the dryer, move a sandwich across the counter – all to no avail. Finally, Dad comes home in a Passat. Junior Darth starts with the Jedi gestures to get the car to do something. Dad uses remote start from inside the house to fire up the ignition. His son is delighted: he got the Force to work. He thinks he started the car with Jedi mind tricks. The ad was praised highly for presenting an automaker that does not take itself too seriously and creates a fun image at the same time – for the first time, it seemed that Volkswagen had touched a chord among viewers of the Super Bowl and stoked interest in the Passat. *USA Today* reports that the ad took the Number 3 spot in their 23rd annual exclusive Ad Meter. The commercial tied for the highest finish ever by a car ad in the Super Bowl, with a Nissan ad shown in 1997.

In Washington, on the other hand, the politically savvy company is seeking government support. When VW succeeded in hiring Toyota's chief lobbyist, this represented a minor victory in the race for the American market. Some may even interpret this switch as the first winds of a sea-change about to sweep the US market: Toyota, for years the most successful foreign automotive company on the American market, is reeling from the triple whammy of the 2008–2009 sales standstill, its high-profile quality issues, and now the Fukushima catastrophe. Yet most observers would argue that, whilst Toyota is indeed facing more difficulties than ever in its most important national market, the Japanese manufacturer can in no way be written off. Volkswagen certainly stands to make windfall gains in the face of its rival's sudden weakness, gains that may even give it some of the momentum it needs to really do the numbers in the States, but Toyota still understands the American market better than Volkswagen and is simply better established there: although Toyota lost 1.8 percent market share between 2009 and 2010, it still sold 15.2 percent of all automobiles bought in the USA in 2010. Meanwhile, Volkswagen improved its US market share, but by just 0.2 percent to 2.2 percent in 2010.

Volkswagen must find a way to mobilize its US business in a market that the auto industry in general considers an especially difficult one. Toyota and Lexus have demonstrated how this is done: in the USA, sales success is achieved by very successful salespeople. Lexus, for example, has exceptionally few dealers, but very successful ones. The Japanese sell well over 1000 vehicles per dealership. These are figures that Volkswagen can only dream of. Rebuilding a dealership network is a very tough process, since the network consists of hundreds of independent entrepreneurs. This requires a lot of attentiveness and patience – as well as a clear strategy for convincing dealers to make a long-term investment in a joint business. Volkswagen is in something of a bind: it will have to show sustained growth to get dealers on side, but only dealers will get it sustained growth. This is one reason why 2011 is such a crucial year for the Group in terms of sales figures.

Moreover, if Volkswagen really wants to meet its objective of selling a higher volume of cars, it cannot succeed without a major contribution by Audi. The only way for Audi to accomplish this while covering its costs is for the premium subsidiary to finally decide to produce in the USA. This will be a significant challenge for the offshoot, since Audi has no experience of operating overseas factories to date. Yet, without an American plant, the company will always be in at a disadvantage compared to Mercedes and BMW with their US production sites. Furthermore, Audi would have to accept considerable risks in view of a highly volatile euro–dollar exchange rate, with a strong euro always making exports to the USA less profitable. Yet again, despite its high-profile PR successes in America in 2011, the Group is still in a dilemma: to sell more cars in the USA, it has to produce there in order to keep costs down; but producing there only makes commercial sense once it is selling more cars. Chattanooga, with the billions invested there, is the answer to this riddle, running on the old American adage that "you've gotta spend money to make money." In political terms, this is a savvy motto, winning as it does the support of the US government, and large-scale investment in local communities is rewarded with instant loyalty from them. This tangible change in strategy shows just how much of a "traditionally American" company Volkswagen has in some respects become, but it remains to be seen whether this is enough to really succeed in the States.

2.4 POWERFUL COMPETITORS

Taking annual net revenues as a measure, in 2010 VW ranked second among carmakers worldwide, with a revenue of US$183 billion. However, taking annual sales, US competitor General Motors and long-time industry leader Toyota lead over VW – the latter by roughly 1.2 million vehicles in 2010.

After the Great East Japan Earthquake, however, it is widely expected in the industry that Toyota might lose the number one spot, maybe for some years to come. This does not mean that Toyota is no longer one of the most powerful players in automotive industry: the Japanese firm is a definite leader in green car technology, most of all in hybrids; but the carmaker has to stomach the damages of the earthquake and tsunami disaster, and may have to step down from the top spot for some time.

If VW manages to surpass Toyota in 2011, it will be in no small part due to the earthquake. Yet it was the 2008–2009 crisis in the automotive industry that helped Volkswagen in its quest to reduce the gap between it and its rival to the point where it now stands to overtake it. The changed business dynamics during the crisis, the toughest in the 130 year history of the automotive industry, shed new light on the German company's ambitious plans in the race to the top of the automotive industry. When Winterkorn laid down the gauntlet – shortly after taking charge as Chairman of the Board of Volkswagen in 2007 – many of his managers considered his goal of surpassing Toyota by the year 2018 to be positively Herculean. The gap between VW and the Japanese car maker appeared almost insurmountable, considering that Toyota held a sales lead of over three million vehicles: now that lead has been cut by more than half, and the gap looks not just bridgeable, but perhaps even jumpable.

Key to Volkswagen's ambitions is the expected growth in new markets. The Group's volume-based market share in emerging countries is planned to rise from 47 percent to 52 percent by 2018. During the period from 2009 to 2018, the Group expects the automotive sector in emerging markets to reach an overall growth of 60 percent, which some now consider a rather conservative estimate. In comparison, over the same period the established markets of Western Europe, USA, Canada, Japan, Australia, and New Zealand are expected to increase by 27 percent, with a large share of this growth taking place in North America. China is a special case. In 2009, China overtook the USA as the largest car market in the world with vehicle sales of 13.6 million, 10.3 million of them passenger cars. In 2010, VW hit the landmark figure of two million vehicles sold in the People's Republic. In view of this growth potential, Winterkorn is aiming to increase VW sales in China to three million vehicles over the next couple of years.

2.5 CHINA'S STATE-CONTROLLED ECONOMIC BOOM

For Volkswagen, China is the key to the top global position in automotive sales and is therefore well worth considerable efforts. Just one day after the annual shareholders' meeting in 2010, at which Martin Winterkorn had to field questions by critical investors for hours, the VW boss hurried to Beijing to make an official presentation of VW's concept car, an electric version of the Lavida, developed especially for the People's Republic.

The Beijing Motor Show is becoming increasingly important to international manufacturers, not just for the Group from Germany. The standards local consumers have come to expect in a car are rising constantly, and most OEMs have dedicatedly buffed up their motor show appearances – including launches of more and more new models that cater to Chinese customers in particular.

Furthermore, Winterkorn had come to Beijing to outline his company's long-term strategy for all-electric cars. "Our goal is market leadership in electric mobility by the year 2018," the head of VW officially declared. This statement did not emerge by accident, for in the medium and long term, China is the most important global market for Volkswagen and for electric cars in general. Electric vehicles became a hot topic in 2009 when China's auto industry faced mounting pressure to cut down on fuel consumption because of increasing air pollution and rising gas prices. Since then, a phalanx of both foreign and home-grown carmakers (including BMW, Mercedes, Nissan, Dongfeng, BYD, Chery, and Geely) have started to develop plans to launch electric vehicles (EV) in the People's Republic. Volkswagen is eagerly gearing up in this race to next generation cars as China becomes an increasingly important pillar in its overall strategy for battery-driven mobility – and the country's growing megacities are the perfect testing ground for EVs from Europe's largest carmaker.

More than 100 cities in China are inhabited by at least a million people, and the number will almost certainly be growing during this decade. In May 2011 the Republic, which is the world's biggest oil importer, voiced plans to raise annual production capacity of alternative-energy vehicles to 1,000,000 by 2015 as part of efforts to cut oil imports and curb pollution in its rapidly expanding cities – reason enough for authorities to launch a pilot program aimed at weaning consumers away from fossil fuel-powered vehicles. Five cities – Shanghai, Changchun, Shenzhen, Hangzhou, and Hefei – have already been officially declared pilot cities in which authorities offer substantial subsidies for the purchase of green cars and finance the construction of electric charging infrastructure. The development of all-electric and other new-energy vehicles is also backed by local governments at various levels. Owners of purely electric cars in the five pilot cities, for example, have been eligible for a subsidy of CN¥60,000 (€6100 or US$8800) since June 2010, while owners of plug-in hybrid cars can receive subsidies of up to CN¥50,000 (€5100 or US$7300). The government is planning for these schemes to put 100,000 electric cars on the road by the end of 2012, an ambitious number that many observers see as beyond the reach even of a dynamic and yet centrally-planned economy such as China. Yet, speaking in April 2011, the Chinese Minister of Technology Wan Gang (a former Audi engineer) claimed that there were already 14,000 EVs driving in 25 Chinese cities. What cannot be denied is that the Chinese government wants e-mobility, and is willing to spend money to get it.

It is a message that Volkswagen is reading loud and clear. In 2011 the Group, is rolling out a zero-emission demonstration fleet of Golf and Lavida Blue Motion cars for test drives in the nation's major cities, making use of the possibility to gain experience of EVs with relatively low investment. Moreover, the Group plans to invite research institutions, government agencies, the media, and customers to experience its EV technologies. Feedback from test drives will help the carmaker and its Chinese partners develop the most suitable EV models for the market. Domestic development, procurement, and production should make the electric cars "affordable to Chinese consumers," as VW proclaims, and all of this on a tight timetable, too: in 2013–14, the first locally produced electric cars from Volkswagen should roll off the assembly lines of both joint ventures. In early 2011, VW has also introduced the European-built hybrid variant of the Touareg SUV to the Chinese market. The next model in the Chinese lineup to be equipped with hybrid powertrain technology is the Lavida.

Volkswagen's Czech brand Skoda – which is popular in China and must not trail behind – will initiate the production of a Skoda electric car in China in the near future. Skoda has been building cars there since 2006, and now at least 80 percent of components for the Fabia, Octavia, and Superb models are sourced domestically. One of the roles Skoda now plays in the Wolfsburg-based brand group is to fend off attacks by Korean competitor Hyundai, which has become the fastest growing rival of the VW Group in the Chinese market with a growing market share and 702,000 cars sold in 2010. By 2012 Hyundai's third Chinese plant will boost its capacity in the world's largest auto market by 400,000 vehicles to 1 million annually.

In view of the growing significance of electric cars, it is fitting that Volkswagen's former electric drivetrain specialist – and one-time boss of the German automotive supplier Continental – Karl-Thomas Neumann is now chief executive of Volkswagen Group China. The appointment of Neumann certainly emphasizes Volkswagen's intention to take the lead in China's EV market. Neumann would have preferred to continue to perform his official role as chief representative for the Group's e-car strategy from Beijing, but Winterkorn opposed the idea. Later, Rudolf Krebs, former Director of the VW Engine Plant in Salzgitter, replaced Neumann as the man in charge of e-car technology. Nonetheless, the concurrence of expertise in electric cars and responsibility for the Chinese market certainly remains full of promise for the future, since the sprawling metropolitan areas of the People's Republic will play a central role in the advent of the electric era.

Furthermore, it is no coincidence that responsibility for Volkswagen's most important market has become the chosen testing ground for Neumann, who counts as a leading contender for successor to VW boss Martin Winterkorn, along with Skoda boss Winfried Vahland, Audi head Rupert Stadler, and the relatively new sales leader Christian Klingler.

However, the complex structure of the Chinese automotive industry, which forces international manufacturers into partnerships with government operations, makes it difficult for foreign carmakers to implement successful marketing strategies, particularly since Chinese partners have their own brands and also conduct joint ventures with other foreign car makers. In the case of VW's partners, First Auto Works (FAW) has operations with the Japanese car maker Mazda and, through a subsidiary company, even with Volkswagen's arch rival Toyota. Shanghai Automotive Industrial Corp (SAIC), on the other hand, runs joint ventures with General Motors.

Winterkorn is well aware of the importance of this huge and intricately structured market for the success of his 2018 plan, and continually refers to China as "our second domestic market." And, indeed, growth rates are enormous, with experts predicting an average annual growth of roughly 10 percent. PA Consulting Group estimates a vehicle sales level of 25 million as early as 2015, by which stage more than one in every four new cars or commercial vehicles in the world would be sold in China. This forecast is based on a trend analysis of the number of Chinese holders of a driver's license, expansion in infrastructure, and future availability of consumer credit. Volkswagen takes a more conservative view and anticipates more than 60 percent growth in the Chinese car market to over 23 million vehicles (including Hong Kong) per year by 2018 (Figure 2.1).

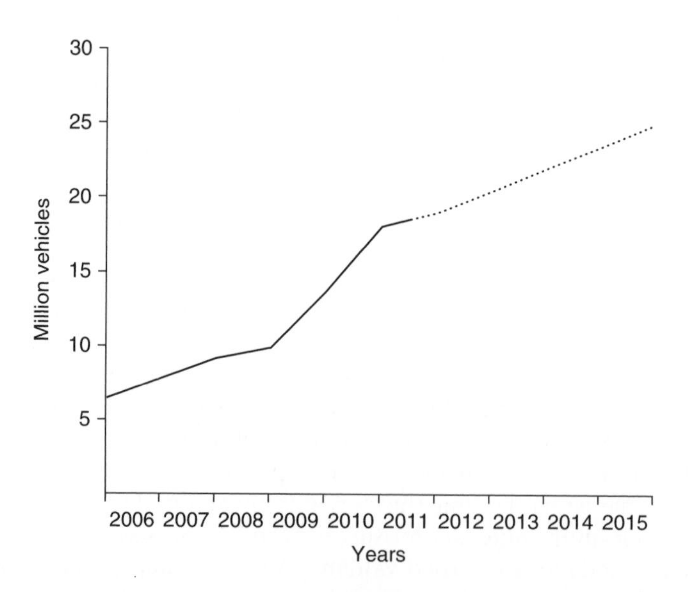

FIGURE 2.1 **Sales forecast for China through 2015 (all vehicle segments).**

Source: 2005–08 figures taken from *Automotive News Europe* (Global Market Data Books) and MarkLines; all figures after 2009 based on expert estimates and modeling by PA Consulting.

Still, the political will of the central planning committees of the Chinese government must not be underestimated and probably is the most difficult factor to anticipate in any evaluation of future market growth. The Beijing municipal government, for example, shocked foreign carmakers in late 2010 with its announcement that, among other measures to help ease Beijing's worsening traffic congestion, it would limit the number of new car and minivan license plates to 240,000 in 2011, just a third of the 2010 issuance.

The appetite for cars in China is still in an early phase given the very rapid explosion of car demand since 2005, and though China has not exactly been immune to a global financial crisis with its roots in profligate Western consumerism, its economy has been much less exposed than many others. During the global economic slowdown, the Chinese government has spent widely on anything from railways to vouchers handed out to buy things like Chinese-made cars and white goods. Unlike many other economies around the world, such as the UK or the USA, the Chinese government has deep reserves and could afford to keep this up for a while. Economic growth in 2009 reached roughly 9 percent and did not slow in 2010; in 2011 it has shown no signs of doing so either. In fact, as China's GDP and inflation surge, experts are becoming increasingly worried about the economy overheating.

China also raised interest rates three times in the second half of 2010, and then again in Spring 2011, ordering banks to issue fewer loans in an attempt to make sure it can meet a 2011 inflation target of 4 percent. Whether these efforts to tighten monetary policy will rein in inflation and keep the economy relatively stable, remains to be seen. With increasing inflationary pressures throughout 2011, the government has now set an annual growth target of 7 percent to ensure sustainable development during its twelfth Five-Year Plan. Most economists, however, expect growth to be around 9 percent in 2011, and slightly lower in 2012. It is hard to predict whether and to what extent the government's measures will affect consumers' appetite to buy new cars. So far, car demand in China has mainly come from the affluent upper class, which comprises about 300 million people. Many local observers estimate, however, that there is now a structural buildup of demand as wealth from China's rapid economic growth over the last ten years continues to percolate through the economy. Even as sales figures stopped growing quite so rampantly in the second half of the year, domestic demand for cars strengthened tangibly overall in 2010 as wealth spread to smaller town and cities, and this effect is likely to keep automakers occupied for some time satisfying domestic demand.

Chinese car owners are relatively few in number and still share the streets with well over 500 million traditional bicycles. With an overall population figure of 1.3 billion, there are currently about five cars for every 100 people in China, whereas there are as many as 50 cars on the road for every 100 Germans or 85 per 100 Americans. Although the comparison to North

America is certainly not realistic, the pent-up demand for automobiles in China is undoubtedly enormous. It is figures like these that set automotive managers fantasizing with their promise of long-term growth and make confident of the future even during slight growth slow-downs such the one that has been experienced in the early part of 2011.

With so many drivers being new to car ownership (estimated at 80 percent of new car customers), the novelty will take time to wear off. The appetite to get on the car ownership ladder is still very strong: it is a lifestyle and status issue that transcends other considerations, especially in a period of such rapid change and economic growth. This trend may alter with growing car ownership, but for now China's population is enjoying something of a honeymoon period in its relationship with the automobile. And if sales in the established inner cities are slowing relatively quickly, there seem to be plenty of new buyers in the suburbs and further out in the smaller cities and towns.

Moreover, times are long past in which the Chinese were satisfied with previous generations' technology: customers in the communist country have increasingly become status-conscious and are demanding nothing less than what is considered the latest and greatest in Europe, Japan, and the USA. OEMs like Volkswagen are well advised to launch technological innovations promptly in this market. In 2007, Volkswagen Group China introduced two of its core Blue Motion technologies – the TSI engine and DSG transmission. The TSI (turbocharged stratified injection) engine enables highly efficient combustion of fuel while the DSG (or double-shift gearbox) transfers power smoothly and efficiently for optimal use of energy. The carmaker now has TSI engine production sites in Dalian, Shanghai, and Changchun, as well as a DSG plant in Dalian. Volkswagen's China-made models with TSI engines and DSG transmissions include the Lavida, Magotan, Sagitar, and Skoda Octavia Mingrui. In the first half of 2010, the Lavida was the second-best-selling car in the People's Republic after the F3 manufactured by the privately owned Chinese carmaker BYD.

In May 2010, the German carmaker even announced that it had reached the target set in 2007 to cut fuel consumption and emissions in its China-made fleet by 20 percent, mainly due to its TSI engines and DSG transmissions (Table 2.2).

Both VW and its Chinese competitors benefited from the incentive package announced by the government in December 2008. It included a purchase tax cut of 5 to 10 percent for cars with engines smaller than 1.6 liters (97.6 cubic inches), and a scrappage incentive that gave one-off cash rebates totaling an estimated US$730 million (€507 million) to owners who traded in older vehicles for newer, more fuel-efficient ones.

The halving of the sales tax for small cars and special support for car buyers in non-urban areas have resulted in higher demand, though the importance of the incentives seems to have been more as a psychological adjunct to the

TABLE 2.2 **Car sales of the ten largest OEMs in China, January–July 2009 compared to January–July 2010 (in 1000 units; rates of change and market share in percent)**

Passenger car sales by manufacturer (in 1000 units; market share and change rates in percent)

Manufacturer	Sales, first six months 2009	Sales, first six months 2010	Change	Market share 2010
Shanghai – GM	264,800	440,600	66,4	9.6
Shanghai – VW	304,400	415,800	36,6	9.1
FAW–VW	306,200	401,600	31.2	8.8
BYD	176,800	289,000	63.5	6.3
Beijing – Hyundai	235,600	285,400	21.1	6.3
Dongfeng – Nissan	198,400	284,000	43.1	6.2
Chery Automobile	180,000	223,700	24.3	4.9
Geely Automobile	147,000	201,000	36.7	4.4
Chang´an Ford	138,400	197,400	42.6	4.3
FAW–Toyota	142,400	185,600	30.3	4.1
Total*	2,094,000	2,924,100	39.6	64.0

* Total sales of the ten largest manufacturers.
Source: *Germany Trade and Invest*, "Chinese Car Market's momentum slows down", September 21, 2010.

broader stimulus package and what it said about the authorities' attitude to economic growth. The stimulus measures served to quickly overcome initial caution on the part of consumers in early 2009 as the international financial crisis unfolded. When it became clear that the authorities would act quickly to prevent economic slowdown in China, confidence returned, consumers were reassured, and the interwoven underlying drivers of rising car demand in China were in the ascendancy again.

Local analysts assign more importance to China's economic advance and the factors associated with that than to the stimulus measures, though they acknowledge that lower taxes have helped as a part of the mix. However, the unprecedented surge in demand for cars is more a reflection of growing real incomes, a rapidly rising urban middle class, continuing urbanization, infrastructure development, and rising asset prices (real estate and stock prices). All the while, with rising incomes more people are moving into the annual income territory (variously estimated at around US$4500; €3100) in which car purchase becomes economically feasible.

In late 2010, the Chinese government correspondingly announced that, to slow down the now overheated market, the tax on cars displacing 1.6 liters or fewer (which had already been reduced) would be phased out. As a consequence, consumers had to pay at least CN¥2317 more (US$340 or €235) if buying a ¥100,000 (US$14,640 or €10,150) car with 1.6-liter or lower displacement. Separately, the Chinese government's "cash for clunkers" program, providing subsidies ranging from ¥3000 (US$440 or €305) to ¥18,000 (US$2640 or €1830) per person for replacing their old, polluting vehicles with new ones, was abandoned in 2011.

In 2010, VW earnings in China amounted to €4.3 billion and the proportionate operating profit was €1.9 billion. These figures for the joint venture companies in China are not included in the Group's operating results, because these companies are accounted for using the equity method and their profits are exclusively allocated to the Group's financial results on a pro rata basis. Total profits before taxes in 2010 were €8.994 billion, a respectable €1.892 billion coming from the Chinese joint ventures. Another figure shows just how important China is for the well-being of the Group from Wolfsburg: without the healthy gains in the market, 2009 overall sales would have declined by 6.1 percent.

The Germans skim profit from their China business in three ways. First, VW participates in profits from its joint ventures: if the revenues exceed a defined minimum amount of liquidity, the surplus is either distributed to the respective shareholders or reinvested. The dividends VW obtains are based on its stakes, which are 50 percent in the case of the joint venture with SAIC, and 40 percent in the joint venture with FAW. Second, VW ships knockdown kits to China which are assembled locally. With this strategy, Audi still manages to achieve a "local content" – i.e. parts produced in China – of 50 to 60 percent per vehicle. Still, European plants supply components such as engines, which are then purchased by the Chinese joint ventures – also a source of profit for VW. Finally, Volkswagen ships completely finished vehicles to China, such as the luxury sedan produced in Dresden, the Phaeton, and the Touareg, which is built in the Slovakian city of Bratislava. Winterkorn would like to further boost exports to China of these high-margin vehicles, but the Chinese sales organizations prefer locally manufactured products.

Volkswagen's Chinese subsidiaries have their own sales organizations, and with the extremely high market growth of recent years, their returns are high. Just as in any other sales region, Chinese car buyers have come to expect discounts on catalog prices. Driven by the boom, however, prices remained pretty stable all through 2010, with a moderate market-average drop of 3 percent; even with the slight slow-down in growth setting in 2011, this is still very much a seller's market from which Volkswagen stands to make healthy margins.

Of all things, it was Volkswagen's affinity to politics and its representatives – which in Germany had repeatedly been criticized as a burden – that has proven helpful in dealing with the Communist government in China. As early as 1978, when the Communist leadership first set its course for opening the country and its still fledgling economy, the first Chinese delegation came to visit Wolfsburg. Long negotiations – interrupted by the second oil crisis – followed and eventually resulted in the founding of Shanghai Volkswagen Automotive Company in 1984, the very first joint venture for automotive production. Its first model, the VW Santana, soon established a visual presence on local streets, first as a standard vehicle in government authorities' fleets and later as a taxi. The founding of FAW-Volkswagen Automotive Corporation in Changchun followed in 1990.

Even today, one of the most challenging tasks VW managers are facing in China is to nurture the sensitive network of relationships with its joint venture partners and authorities. Right from the start, Volkswagen invested money in Chinese factories producing transmissions and engines, as well as local development centers. Since its market entry in 1984, the German car maker has created at least 33,000 direct and over 300,000 indirect jobs, and has sold more than 10 million cars. Relations with Chinese officials are excellent and the signals for VW have been so clear that the Supervisory Board of the carmaker saw reason to increase its investments in China by €1.6 billion (US$2.3 billion).

To date, Volkswagen has primarily been strong in northern China, while the Japanese rival holds the lead in the South. Henceforth, a substantial share of VW investments is to flow to the south of the country in order to build up new capacities there. This money should serve to defend the company's leading position, since its market share has begun to shrink in light of the steadily increasing competition from international and domestic carmakers. While Volkswagen was still able to claim 18.7 percent of the market in 2008, in 2010 the VW share was 16.8 percent. The funds for investments defending VW's leading position in China are taken directly from the cash flows of the joint ventures SAIC and FAW-VW.

Generally, the Chinese government tends to be merciless with foreign producers: access to the Chinese market is granted only in exchange for knowledge transfer. No foreign manufacturer may produce independently, but must instead enter the market in a joint venture with a domestic partner. This obligation to localize does not apply just to the assembly of vehicles, but also to the manufacture of a number of key components. This is how the Chinese acquire the expertise needed to build crucial parts such as engines and transmissions, and the state cleverly exploits the attractiveness of its large and growing market to import the knowledge required to continue developing its own potent industrial sector.

Foreign carmakers are bowing to pressure from Beijing to transfer more technology to their Chinese partners by launching jointly designed brands aimed at small-car buyers on the mainland, hoping the move will help them compete for buyers at the low end of the Chinese market, which has been the fastest growing, and most competitive segment. "All joint ventures of foreign car manufacturers – like FAW-Volkswagen and SAIC Volkswagen – have been asked to develop local brands," says Neumann, confirming plans to launch a joint brand with the German carmaker's two local partners. VW sees an indigenous brand as "an opportunity to get into new market segments, especially those we have not been able to get into to so far." Since late 2010 the Group has been in talks with its Chinese partners, looking at the market for models below €8000 (US$11,000).

All global carmakers are eager to placate Beijing, and Volkswagen takes painstaking account of the sensitivities of its Chinese business relations. In fact, such is the company's political talent in dealing with its Chinese partners that it has twice come in for heavy international criticism for its closeness to the regime. In 1989, at the time of the bloody suppression of student protests, which elicited anger and sadness worldwide, Carl Hahn, CEO at that time, stood by his Chinese partners and promised them that he would continue to drive automotive development in China regardless of the political conflict. This gesture, which was heavily criticized in Germany, not only increased political acceptance by the Chinese leadership, it also promoted identification of local employees with their then relatively new employer.

Decades later, in the context of the 2008 Olympic Games in Beijing – which were also heavily criticized globally due to the lack of freedom of expression and the oppression in Tibet – Volkswagen was able, with the help of targeted marketing efforts, to become part of the Chinese self-promotion. During the two-week show, the Germans provided the official vehicle fleet, and VW became a national Olympic symbol. Today, the state president, prime minister, and parts of the political party use vehicles from the German brands. In advertising messages, it cleverly appeals to the Chinese tendency toward status thinking: VW paved the way for its premium brand Audi for example with the slogan "For successful people." However, this proximity to the political elite has meanwhile become something of a burden. Many individual customers and private entrepreneurs prefer to distance themselves from the tastes of the political cadres in China and have started to go for BMW and other luxury brands.

Meanwhile, the Group in China offers all 50 VW models from its product lineup, 20 of which it already produces locally. The two joint ventures manufacture the VW Polo in several versions, the Lavida, Santana, Santana Vista, Passat New Lingyu, Touran, Tiguan LWB (a version with extended wheelbase), several variants of the Golf, New Bora, Jetta, Sagitar, and Magotan, and the Skoda Fabia, Octavia, and Superb. Audi, present in the country since 1990, builds the A4 and A6 models with extended wheelbases as well as the Q5 SUV. Seven new models are to be added to this range in the near future. In the early years, the Wolfsburg car maker only produced special editions for China, as facelifts of models that had already run the course of their production lives elsewhere, but since 1999 brand-new VW models such as the Audi A6, the Bora (2001), the Bora HS and later the New Bora have been rolling off Chinese assembly lines at the same time as in Europe.

To survive the increasingly tough competition, car makers in China must continually lower their costs. In 2000, VW still had a market share of 53.2 percent, but at the time former China chief Vahland took over in 2005 this had sunk to 17.3 percent. One reason for this was intensive rivalry between the two joint ventures. Vahland, previously CFO at Skoda, pushed down costs by over 40 percent, increased production output, and reduced the number of platforms upon which models built in China are based.

Although product planning and control as well as the styling of the models are determined in Wolfsburg, VW takes care to involve the Chinese to an extensive degree. Above all, Volkswagen utilizes local partners' specialist knowledge in developing cost-effective components. VW pushes the share of locally purchased components per car, with local sourcing at more than 85 percent. In the case of new models such as the Lavida or the new Bora, the figure is nearly 95 percent; just six years ago, the degree of localization was still languishing at 60 percent. Still, the cost reduction during the Vahland years does not yet represent the maximum savings potential, and as the explosive growth in the Chinese market in 2009 and 2010 slows to less outstanding levels in 2011, cutting costs will become more and more pressing if the financial results of Volkswagen's Chinese operations are to hold up.

In terms of products and technology, the Group orients itself toward Chinese needs. This includes offering a larger share of visible chrome features for status-conscious drivers, as well as coatings developed specifically for the acid rain showers that frequently occur in Southern China and various market-specific modifications to the vehicles: "We have been listening very precisely to what the Chinese want. Since they weigh several kilograms less on average, we designed the seats to be softer. In addition, customers want to get on the car without bending down, so we make the cars longer, even the Tiguan," CEO Winterkorn explained on this matter.

The strategists at sports car manufacturer Porsche, meanwhile, know exactly where – as a new brand in the VW stable – they can make money and gain market share in the coming years: the four-seat Porsche Panamera was not presented in Frankfurt or Detroit in 2009, as usual, but straight away at the motor show in Shanghai. VW managers estimate that by the year 2015, China might have developed into the largest premium market in the world. The country is already expected to be the biggest sales region for the VW Phaeton and Audi A8 luxury sedans; in 2010 about every third Phaeton went to China, despite the luxury tax levied there. Volkswagen will most likely sell 3000 units per year in China, so it is no wonder that the new model of the luxury sedan, which is difficult to sell in Europe, was presented straight off the bat in China in June 2010.

However, Volkswagen had been tenaciously refusing to consider the other end of the automotive scale: it had no feasible strategy for entry-level or low-cost models. To date, domestic producers such as Hafei, Great Wall Motor Company, and Geely are doing especially well in this crucial segment, offering low prices and intensifying rivalry in the already fiercely competitive Chinese market for small cars. Just like most other foreign brands in China, VW enjoys a reputation for quality that could be jeopardized if the carmaker goes too far down-market with its global brands. The new brand to be launched in the People's Republic could thus prove helpful for both sides: VW, which lacked an entry-level model, and the Chinese partners, which hope for a closer knowledge transfer.

China's growing manufacturers: Challenging the West

Since its entry into the World Trade Organization (WTO) in 2001, China has been following a course of relentless expansion. This growth had been planned for a long time, since the country's economic strategists know that modern industry does not work without mass transportation. But the plan goes further: the powerful elites in China do not want the country to be merely a gigantic sales market for the global automotive industry; rather, their plan is to lead the industry globally with independent and progressive Chinese manufacturers in the future. It is the avowed goal of the state's Commission for Development and Reform to make the automotive industry one of the central pillars of the Chinese economy. Accordingly, China's government still has plans to consolidate its fragmented automotive OEM landscape and focus increasingly on vehicle export.

The Chinese government has long favored consolidation of the industry into bigger groups able to produce at lower unit costs, in an attempt to encourage more extensive research and development and ultimately enable international competitiveness. But that objective has been hampered by the industry's regionally based structure and the support that the provinces give to their local and highly vertically integrated automotive corporations. Some of the largest Chinese manufacturers, such as SAIC, FAW, Dongfeng, and Changan, already have production capacities of around two million cars each (or higher), and the government has given them the official mission to swallow up smaller producers in the course of the planned market consolidation. The most recent five-year plan for the automotive industry provides for building up strongly competitive mixed corporate groups around these four companies. The goal is to create two to three car producers with an annual production of over two million vehicles and another four or five manufacturers producing over one million units. An increasing share of production – up to 20 percent – is to be earmarked for export.

Thus far, the absorption of Nanjing Auto into SAIC is a rare example of consolidation among the bigger groups. Central government has not been able to effect change: since its first call for consolidation, not much has happened. Market forces may help to bring about more consolidation as the big firms start to enjoy greater economies of scale and higher sales from national distribution. The future for the smaller independents who have grown out of nowhere over the last five years – the likes of Great Wall, Chery, Geely, and BYD – is uncertain if a shakeout to the industry starts to occur. If it does, that shakeout could also become the battleground for a political struggle between Beijing and the provinces, with the regulatory playing field also assuming greater importance in the area of environmental incentives and, possibly, incentives for electric cars – some companies being better placed than others in that area. Even in terms of technology, the leadership in Beijing is dictating the broader terms. The goal is to capitalize on green car technologies in order to create a competitive advantage for Chinese producers even in the established markets of the "triad" countries (Europe, the USA, Japan).

In the field of conventional engines, the Chinese have long accepted that they cannot close the gap between themselves and American, European, and Japanese engineers at the drop of a hat; these countries have up to a century's experience in conventional, gas- and diesel-driven powertrains. In the relatively new field of battery-driven cars, however, Chinese engineers see themselves on a par with Western competitors. China has long become the world's global electronics workshop, producing technologically complicated products such as advanced batteries. Like the Japanese companies that dominated the field of consumer electronics in the late 1980s after many years of strategic investment, or South Korean companies that took the lead in the production of flat-screen monitors in the 1990s, the Chinese aim to claim the top spot in the production of electric vehicles before the year 2020. For over ten years now, China has committed leading universities and institutes to this strategy and has pumped billions of yuan into the development of high-performance batteries and drivetrain technologies. Chinese manufacturers are obliged to offer at least one green car model in their new product lineups, supported by extremely low-interest government loans.

In mid-2010, the government announced its plans to invest more than CN¥100 billion (€10.15 billion or US$14.64 billion) in alternative-energy vehicles during the coming ten years. The funds are intended for the development of both battery-driven and hybrid vehicles, aiming for at least 1 million on the roads by 2015. The Chinese government also wants to push the research and development of fuel-cell cars. The fleet of green cars on Chinese streets (plug-in hybrids, battery-driven and fuel-cell cars) is planned to reach the 5 million mark in 2020. By the same date, annual sales of green cars of all sorts are scheduled to break the 3 million mark. 2011 has seen top level

Chinese officials state and re-state these ambitious goals, writing them into the new Five-Year Plan.

When it comes to electric infrastructure, the government is also pressing ahead: the state-controlled power utility State Grid Corporation of China has already installed charging stations for electric vehicles in Beijing and Shanghai. These facilities are urgently needed, because hardly any residents in Chinese coastal cities have a private garage that could be used to charge vehicle batteries. The Chinese can already point to a success story related to electric driving: an electrically powered bicycle which typically reaches a speed of 20 kilometers per hour (12.5 miles per hour), has a range of up to 50 kilometers (31 miles) and requires between 4 and 8 hours to charge. Over 100 million Chinese already use these, and at least 20 million more are expected annually. An electric bike with a lead battery costs the equivalent of between US$190 and 325 (€132–225). The more advanced variant with lithium-ion batteries is more than twice as expensive, but also offers twice the driving range, according to manufacturers' data. The competition is fierce: over 2700 licensed producers and countless smaller ones are crowding the market, building over 20 million units annually in total. The same customers could purchase their first electric car in five to ten years' time, and much speaks in favor of this originating from domestic production as well.

Fittingly enough, some of the world's largest suppliers of lithium-ion batteries are located on the Chinese mainland and are working on this type of technology as an international standard. Joint ventures and alliances with Western companies are also high on the agenda. One supplier of green car battery technology that should be taken very seriously, for example, is the joint venture Advanced Traction Battery Systems (ATBS) founded by SAIC and the US company A123 Systems – the first alliance to be established between a non-Chinese battery manufacturer and a leading Chinese automaker. Headquartered in Massachusetts and founded in 2001, A123 Systems' proprietary nanoscale electrode technology is built on initial developments from the Massachusetts Institute of Technology. The joint venture with the Chinese focuses on the development, production, and sale of complete, battery-powered drive systems for hybrid and electric cars as well as for trucks and buses. In addition to developing battery packs for the new 2012 model-year electric passenger car from SAIC, A123 currently supplies battery technology to SAIC for several of its electric drivetrain vehicles.

As early as 2009, SAIC announced plans to invest CN¥12 billion (US$1.76 billion, €1.22 billion) within three to five years in developing new-energy sedans and auto parts. At the end of 2010 China's largest automaker launched the self-made Roewe 750 mild hybrid, offering 20 percent fuel saving, as well as the Roewe 550 plug-in hybrid. An all-electric A00 hatchback is supposed to follow in 2012.

Nonetheless, analysts doubt whether the successes of green automotive technology in China will be achieved at such a rapid or sustained pace. So far, the state has been the biggest customer for hybrid and electric cars, and hardly any of the vehicles end up in the hands of private customers. For example, the Prius hybrid model from Toyota has a big market in Japan – more than 300,000 vehicles of this type were sold there in 2010 – but in China the sales volume has been below 4000 units for the past three years, and in the first half of 2010 the Japanese hybrid flagship was virtually unsellable.

Chinese electric dreams: BYD

The most stellar example of a Chinese automobile business carries a very positive-sounding name: Build-Your-Dreams (BYD). Founded in 1995 in the southern city of Guangdong to manufacture rechargeable batteries, in 2008 it had a 30 percent share in the global market for cellphone batteries and over 60 percent market share for nickel-cadmium batteries used in laptops. By its own data, BYD already employed over 200,000 people in 11 Chinese industrial parks in 2010.

Since 2003, its president and founder Wang Chuanfu has been counting on the automotive industry: BYD was the first private company in China to set up automotive production. In September 2010, BYD announced it had reached the total of one million vehicles sold since its launch only seven years earlier. BYD has maintained a growth rate of over 100 percent for five consecutive years. In 2009, over 448,000 vehicles were sold, and the popular BYD F3 model was the best-selling car that year at 291,000 units. Wang has also changed his brand's visual image: the original company logo looked so much like the BMW emblem that he had to replace it in 2009 to avoid impending legal claims.

In 2010, however, BYD (which has acquired a reputation for some pretty ambitious talk in recent years) missed its Chinese sales target, despite having slashed it by a quarter in August: BYD only sold 519,806 cars, making an increase of 16 percent over the previous year. Nevertheless, the ambitious privately owned manufacturer wants to be the largest car maker in China by 2015, and by 2025 at the latest the company even plans to catch up with the Top five of the international automotive industry. Lying at the core of its expansion is its expertise in lithium-ion batteries as well as electric drives.

The company entered the limelight in the fall of 2008 when US investor Warren Buffett came on board. The financial guru bought a 10 percent stake in the company for a respectable US$232 million (€160 million) – an accolade for BYD. Internationally, there is no longer any doubt that BYD must be taken seriously. The value of Berkshire Hathaway's investment in BYD grew more than tenfold in a little over a year. However, after a series of bad news stories, BYD shares lost 40 percent in 2010.

Volkswagen announced intentions to work with BYD to develop electric drives, and at the 2010 Geneva Motor Show Daimler revealed plans to launch an e-car in the Chinese market jointly with BYD. In March 2010, BYD began selling electric-hybrid vehicles. For 2011, Wang announced plans to launch his first electric vehicles with a driving range of 300 kilometers (186.5 miles) in the USA and Europe. Meanwhile, however, experts are re-evaluating BYD's electric-vehicle strategy. The carmaker may not deliver its E6 model to the USA before the second half of 2011. At the 2006 Beijing Auto Show BYD showed an all-electric car called the F3e which, however, might never be put into production due to the lack of an enabling environment for electric cars in China, as Wang Jianjun, deputy general manager of BYD Automotive Sales, explained publicly. BYD acknowledged it had changed its mind somewhat in 2010 after a market investigation and consultation with dealers. BYD points out that there are still problems with the supporting infrastructure and market environment and wants to put more focus on transitional products. This is why the company has developed the F3DM, a plug-in hybrid compact sedan.

In fact, the Shenzhen-based automaker, which has used the hype about e-cars mainly as an opportunity for self-promotion, did not start mass production of its all-electric E6 in mid-2010 – contrary to earlier announcements. Having been so eagerly engaged in development of all-electric vehicles, BYD said that it sold only 54 electric E6 and 290 F3DM hybrids between January and October 2010, and is now backpedalling. Following the sales figures, it announced that it will only produce 100 models of the E6, aimed at taxi drivers in the economic zone of Shenzhen. However, customers buying the car would not only enjoy CN¥60,000 (€6100 or US$8800), from the state but also be exempted from licensing fees of ¥45,000 (€4600 or US$6600). The BYD hybrid F3 DM is considered not fully competitive because catalog prices are as high as CN¥170,000 (€17200 or US$24,900). In Shenzhen, however, buyers of the hybrid would enjoy a ¥30,000 (€3050 or US$4400) incentive.

In its defense, BYD explained that it was infeasible to extensively promote all-electric vehicles before the necessary supporting infrastructure has been improved in the country and would, for this reason, be shifting focus from all-electric vehicles to plug-in electric vehicles and from private transport to urban public transport. Following this very public announcement from BYD, however, the City of Shenzhen now plans to build 7500 charging piles for all kinds of new-energy vehicles at the end of 2011, rising to 12,750 by 2012. Additionally, the Chinese government will grant subsidies to individual buyers of new-energy vehicles in 12 more pilot cities, the Chinese vice president of the Ministry of Science and Technology (MST), explained in late 2010.

While the development of new-energy vehicles in China had been hotly anticipated, by late 2010, after having spent millions of yuan in this field automakers were disappointed with the sales figures of green cars. The

Chinese government has responded in 2011 by reiterating its wish to boost the number of next-generation cars with fuel saving and eco-friendly engines on Chinese streets to half a million by 2012 – but this is still a long way off. According to research company Dratio, which carried out a survey in 2010, 89 percent of Chinese consumers were not interested in purchasing new-energy vehicles due to their high price and the lack of supporting facilities.

Car makers domestic and foreign point out that though the subsidies for electric vehicles are high, the subsidies for hybrid cars are often only CN¥3000 (US$440 or €300) – and this has led to plummeting sales of hybrid vehicles. VW sees the prospects of hybrids in China less positively than rival Toyota and is focussing more on e-cars instead. "There is a role for hybrids in China, but nowhere near of the same size as, say, in the USA," China chief Karl-Thomas Neumann explained. A lot will depend on state subsidies, however, even with battery-driven cars: "It is a hard task to convince Chinese customers to opt for an electric vehicle: most of them are still purchasing a car for the first time ever and are less willing than Europeans or Americans to take limitations into account."

Toyota, on the other hand, has announced plans to partner with the China Automotive Technology and Research Center (CATARC) in a program which will see the Prius plug-in hybrid put through its paces in China. "As we move to distance cars from petroleum, we view electricity, which can be produced using any primary energy source, as exceptionally promising. Greater use of it is in lockstep with China's electrification policy," said Takeshi Uchiyamada, Toyota vice president announcing the test in October 2010. The car had made its China debut at the Beijing Motor Show.

Toyota: lagging behind in China

For a long time, Volkswagen's Japanese competitor had difficulties in the Chinese market. Observers cite cultural resentments between the two countries – perennial enemies over the course of history – as the reason for this. Toyota's market entry was not decisively accelerated until 2000. Until then, cars had only been exported to China from Japan. Akio Toyoda, who had been responsible for Toyota's Chinese business in the years 2001 and 2002, has made China a high priority since his rise to the top management position, and Toyota has increased its Chinese production capacity to over 820,000 units in 2010. The Chinese automotive yearbook claims that the Japanese market leader plans to sell 1 million units in China by 2012 – that would roughly reflect a 10 percent increase each year, keeping pace with estimated market growth.

After taking over as CEO, the first overseas market Toyoda visited was China. In March 2010, Toyoda flew to Beijing to personally apologize for quality problems and a nationwide recall action. That recall – only the RAV4 SUV was affected – had serious consequences: for the first time, in February 2010, Toyota's joint venture with FAW dropped from the list of the top ten manufacturers in China. The Japanese reacted immediately, offering zero-interest loans for some of their models, reduced prices, and expanded their service packages for the RAV4 SUV. In the first half of 2010, they just managed to squeeze back into the top ten.

Meanwhile, the decreasing popularity of Toyota cars and deteriorating Sino-Japanese relations have hit the company's sales in China. Labor disputes over pay have grown in number, and both Toyota and rival Honda were forced to halt production several times in 2010 after workers at suppliers went on strike. Speaking at Toyota's annual shareholders' meeting in Japan in 2010, Executive Vice President Satoshi Ozawa said the company aimed to improve efficiency at its Chinese manufacturing operations to offset an expected rise in labor costs associated with higher wages. However, Toyota has ramped up investment, with the assembly lines of the newest local factory of FAW-Toyota in Changchun intended to start up as early as 2012 and produce an additional 100,000 vehicles for the market.

The Japanese see great potential for their luxury Lexus brand, as do their German rivals BMW, Daimler, and VW's Audi brand. Nonetheless, it should be noted that Lexus was not introduced to China until 2005, and it is still primarily produced in Japan, so it remains unlikely to prove itself a serious threat to cars made by German OEMs with their local production. In fact, Lexus seems to be something of a second choice for prestige-oriented Chinese customers, selling just 100,000 vehicles in China between February 2005 and September 2009, for example.

In production, Toyota handles key technologies more restrictively than the competition; although the company has also founded a joint research center with Tsinghua University in Beijing, there is a reluctance to build up local knowledge. Above all, the Japanese do not want to share technologies and processes related to their lead in hybrid powertrains. On the other hand, the company does want to win over Chinese customers for its hybrid cars, and is counting on increasingly stricter government emissions regulations to play to its strengths. As early as 2005, the first locally assembled Prius rolled off a line in Changchun, in a factory owned by joint venture partner FAW.

While VW are counting on robust demand in China to bring it to the top of the automotive industry, Toyota is worried that prolonged weak sales there could further hamper its recovery. October 2010 was a particularly poor month during which the global market leader saw its first sales drop in 18 months, although Toyota's Chinese sales were at roughly 850,000 units in 2010. In 2011, Toyota is going to place a major focus on small-displacement vehicles which have so far been driving the big sales figures in the Chinese market. The Toyota

Verso was launched at the Guangzhou Auto Show in 2010. Powered by a 1.6-liter engine, it is the third small-displacement car Toyota has offered in China.

2.6 EXPANSION IN EMERGING ECONOMIES

Volkswagen's focus on emerging countries follows an appealing logic: while established automotive markets were essentially stagnating even in the pre-crisis years with total sales of about 37 million passenger cars between 2000 and 2007, BRIC states grew rapidly. Many indicators point to Brazil, Russia, India, and China increasing their global market share to about one-third in the coming years, while the global market share of the "triad" countries is expected to drop to below half.

Latin American markets continued to grow in 2010: Volkswagen's sales in this region increased by almost 10 percent in the first nine months to more than 900,000 vehicles, while market share in the same period dropped slightly below the 20 percent mark. In their largest regional market (Brazil) in 2010, the Germans sold 727.790 cars, reaching an almost 23 percent market share. Rival Toyota is still lagging behind in the big Latin American car market: in 2010, the Japanese carmaker did post record vehicle sales, fueled by credit expansion, as the company explained. Yet sales only totaled 99,570 vehicles – mostly Corollas – up from just 93,486 units in the previous year.

Meanwhile, continued credit expansion fueled Brazilian car ownership. In 2010, a total of 3.51 million cars and light trucks were purchased, an 11 percent increase on 2009. VW designers and sales managers have successfully tailored VW models like the Fox and Gol subcompact cars to the needs and price expectations of South American customers. Accordingly, the bestselling car in Brazil in 2010 was the popular VW Gol (transl. "door") with 294,000 units.

In the years up to 2018, Wolfsburg market researchers are forecasting a compound annual growth rate of roughly 5 percent for the country. Brazil is the most significant automotive market in South America, growing in 2009 despite the automotive crisis thanks to state incentives. In 2010, Brazil overtook Germany as the fourth-biggest single car market in the world, behind China, the United States, and Japan. With a population of 192 million, the country has approximately one vehicle for every seven residents, leaving plenty of room for growth. However, for some years now, Volkswagen has been outperformed there by its Italian competitor Fiat, mainly because the German group lacked a heavy-duty pickup truck model. In 2010, VW introduced the Amarok, produced in Argentina, which is intended to lure customers away from Fiat. Globally, the niche segment of pickup models accounts for a not inconsiderable 6 percent of the car market.

In fast-growing India, VW is trailing a long way behind the Japanese. However, the Germans plan to make India another boom region for them: by 2018, the Group wants to sell 1 million cars on the subcontinent, by which time the Indian car market is expected to have doubled to about 5 million vehicle sales per year. The Group hopes that its alliance with Suzuki will help to expand its market share in India, because Suzuki owns 54 percent in Maruti Suzuki, the Indian market leader. In 2010, the Joint Venture sold a record of 1,215,087 vehicles on the subcontinent. In comparison, VW sold 32,000 units – with a market share well below the level of awareness (1.1 percent).

For Russia, the Group is estimating that average market growth will be more than 9 percent in the years up to 2018. In 2010, VW sold 133,503 vehicles in Russia, making for a market share of roughly 7 percent. Nevertheless, local production of entire vehicles has been made difficult by a shortage of qualified personnel and gaps in the supply chain. By 2015, VW wants Kaluga to be hitting a localization level of 30 percent – still rather modest compared to other production sites.

2.7 VW PUSHES SALES

A witticism was said to have spread among Wolfsburg managers on the occasion of a vehicle being launched: "We have done so well with our new model that even our sales department won't be able to stop it from selling!" What this reflects is a traditional attitude among top VW engineers that is critical of their colleagues in sales and marketing. For generations, with its complex web of importers, aftermarket sales – the service organization responsible for supplying car parts after vehicle sale – and the company's own dealership operations, the sales department in Wolfsburg has received shabby treatment. Salesmen in the Group are in a tricky position, and it is the engineers that set the tone. Sales managers are gleefully dismissed as automotive lightweights who have a hard time winning customers, even with a technically ingenious product on their hands.

Since late 2009, Christian Klingler, Head of Sales and an appointed member of the Board of Management, has been working with great intensity on markets in which Volkswagen must grow to catch up with Toyota. "This can only happen with a functioning sales organization," Bernd Osterloh (Head of the Works Council) has emphasized. In Central and Eastern Europe, for example, the Group needed to improve its sales strategy. However, the senior employee representative is primarily concerned with the company's global organization: "That is where our weaknesses lie."

This self-assessment does not really match opinion within the industry. In recent decades, VW Group Sales has become a weapon feared by the competition. Time and again, it has successfully launched models on the market that are expensive compared to the competition, and it does this even though the Group is under pressure from its notorious process problems. VW's brand image, price positioning, and profits from the service network and automobile financing have been on an upward trend for years now, except during industry recessions in 2008 and 2009. The German company even performs well in its fleet operator business, and it manages to maintain a profitable balance between high-volume leasing sales and the resale of used vehicles that have been traded in. Now it must find new approaches, such as VW's own used-car brand ("The World Car"), which are intended to complete the marketing loop.

In any event, one thing is clear: the Sales team must continue to transform itself. In former years, all manufacturers needed was to build good cars, but nowadays customers take a more direct approach. Today's car buyers come to the dealership better informed than ever before; they are familiar with the latest technology thanks to the Internet, and they know the best prices and the latest vehicle test results. This has forcibly transformed the way in which cars are sold. Contacts between customers and manufacturers are increasingly taking place via call centers and the Web, i.e. directly to the brand and bypassing the car dealership. Meanwhile, banks – often independent ones – offer financing packages. Service and insurance packages are bundled together, and all of this increases the level of complexity at car dealerships and makes it difficult for them to keep pace with changes.

Klingler has declared a commitment to the "digitalization of our business" in order to get closer to customers. The idea is for the Internet to attract additional buyers to the Group's brands, and here Klingler wants to measure up to "the best of the best," meaning online stores such as Amazon and eBay, from whose strategies car makers can learn a lot about customer service. "With these kinds of tools, we and our dealers can better support customers," Klingler explains. In the future, car owners will be informed by SMS messages as soon as their vehicles have been repaired. To ensure that these expansion plans succeed, the Head of Sales will have to extend dealership networks and digitalize user-friendly processes for the age of BlackBerries and iPhones.

In the fleet business, efforts to intensify customer contacts have already paid off, and the Group is very well situated in the fleet market with its Audi, VW, SEAT, and Skoda brands. Although margins in this business are traditionally thin, in recent years Volkswagen has been gaining ground on domestic rivals such as Daimler and BMW and VW's leasing branch is now the market leader in Germany.

VW is expanding its value creation even further by setting up and purchasing its own large dealerships. In March 2011, VW took over the automobile trading business belonging to the Porsche and Piëch families for €3.3 billion (US$4.752 billion). By selling this entity, one of the World's largest car retailers, the families wanted to generate the cash to finance their part of Porsche's planned rights issue. Starting in 1952, Ferdinand Piëch's mother Louise had doggedly primed the sales branch of the family business for success, and with Porsche's automobile trading business, Volkswagen is gaining access to one of the most successful organizations in European automotive sales. The Group can point to its lean organization as a prime example of restructuring its global sales operations: Porsche Holding Salzburg is described by VW as "Europe's most successful private automobile trading company" and has a particularly strong presence in Austria, Western Europe and South-Eastern Europe as well as China. In the 2010 calendar year, unit sales of 565,000 new and used vehicles generated sales revenue of €12.78 billion. The company employs some 20,900 people.

Klingler himself came from Porsche Holding Salzburg, and he is fully aware of the potential that has been handed to him with the deal. "Under the Volkswagen Group umbrella, PHS will retain its status as an independent organizational unit and continue with its business model unchanged," Klingler said when the deal was completed. "All assets remain intact. That also applies to the automobile trading business relating to all non-Group brands. The controlled growth course of PHS is to continue," he added. Yet this relationship also harbors risks. Among other things, it is questionable whether VW's rivals, such as Renault, will still want to market their cars through the sales company.

2.8 PREFAB VARIETY

CEO Winterkorn is counting on technical standardization of automotive architecture to keep Volkswagen's expansion on course and its slew of new models economically affordable. A system of standardized modules for various models, that is modular toolkits, as well as standardization of components like air-conditioning systems, are further developments in the Group's platform strategy (Figure 2.2).

Stated in simple terms, while Volkswagen's engineers have, until now, been equipping a platform such as the PQ35 with entirely different bodies for the VW models Golf, Golf Plus, Golf Variant, Jetta, Scirocco, New Beetle, Touran, and Tiguan, in the future they will be able to simply combine modules within the modular toolkit system in different ways, similar to Lego blocks. New cars can then be developed more quickly and at a lower cost. Until now, any deviance from the standardized specifications of a given platform has generally entailed high costs; engines and transmissions sometimes became expensive custom developments, as illustrated by the Audi A2 compact model. Its advanced technology, including many

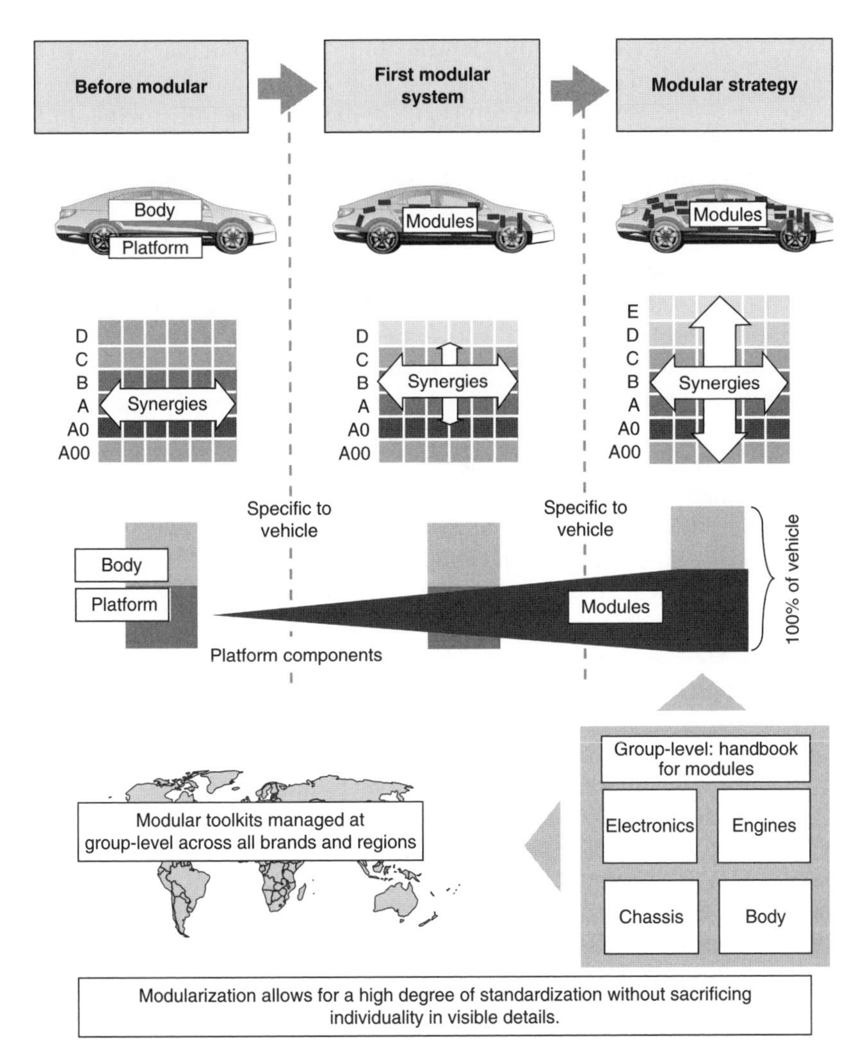

FIGURE 2.2 **Transition from conventional platforms to modular toolkits.**

Source: *Volkswagen – The Integrated Automotive Group*, Presentation by Prof. Dr. Martin Winterkorn and Hans Dieter Pötsch, 2010.

individually engineered components, was highly praised across the board, but commercially it was a flop, so the high R&D costs could not be covered. Modularization should make procurement simpler, and the significantly higher part volumes should make them considerably cheaper, rationalizing production tremendously.

Audi has already exercised this new flexibility in its "modular longitudinal toolkit" (MLB), termed "longitudinal" because, in large models, the engine is not mounted transversely as in compact VW models. The MLB toolkit, introduced by Winterkorn while he was Head of Audi, has created a cost-saving basic architecture for nearly all future models from the A4 onwards, allowing the brand to access market segments it has never been able to break into before. In fact, the first Audi model to be based on the MLB, the A5, is now the basis for an additional product series of its own: initially, an A4 coupé was planned, so that the longitudinal toolkit with new engines, transmission and electronics would not have to be tested on the A4 itself, the most important high-volume model in Audi's Avant (station wagon) or sedan line. Meanwhile, a convertible version and what is referred to as a five-door "sportback" version were derived from the three-door coupé that debuted in 2007 under the independent model name A5.

The modular toolkit system allows the Group to cover niche segments, with Audi filling the gap between its A8 flagship and the A6, for example, with a coupé variant of the A6, including a sportback version, the A7. When one segment loses its appeal, VW fills the next with a new model. The Q series is a classic example of how this works: in 2007, Audi produced over 77,000 units of the large Q7 SUV; however, sales of the Q7 declined to 59,000 in 2008 and to just 28,000 in 2009. Yet its smaller brother, the Q5, has now taken over with more than 105,000 units produced in 2009 after just over 20,000 units in its launch year 2008. With the compact SUV, Audi obviously has caught up with a new model trend, and in the first nine months of 2010 it had already sold more than 109,000 compact SUVs.

Volkswagen's premium branch is consistently extending its model lineup. A small Q3 SUV is to be built by SEAT in Spain in 2011. The Audi A1, which is based on the Polo, is being manufactured in Brussels, Belgium. For 2011, Audi plans to deliver about 80,000 A1 cars to customers, and average annual sales of 100,000 units are being targeted. However, it will be difficult for Audi to gain ground in the small luxury car niche, as rivals such as BMW are traditionally strong there – BMW is ready to defend its position in small, up-market cars with the cult brand Mini. Following the A1, there will be an Audi City based on Volkswagen's new subcompact car, the Up. When Winterkorn became Audi CEO in 2002, this branch of the Group counted only half a dozen model series; with the A1 and the Q3, that figure has been doubled. Using the MLB, developers designed the A5 and A4 models first, then the A7 and A6. The A8, not yet being based on the modular toolkit, in future will very likely be sitting on the new MSB (modular standard toolkit) developed by Porsche.

The Group intends to apply the modular toolkit system used by Audi to the core brand, Volkswagen. Like Audi, which is churning out new models based on the MLB, Wolfsburg wants to follow suit with the Modular Transverse

Toolkit (MQB – see Figure 2.2). The Group's first compact car to utilize the MQB is the Audi A3 in 2011. About a year later, the most important Group model, the VW Golf, will be based on the MQB.

The MQB is designed as a basic chassis architecture with four different wheelbases and six track widths, thus able to accommodate the entire VW line-up in the subcompact and compact class worldwide. Engineers can use the MQB just as well for a sedan or compact car as for an SUV or sports car. Modules reduce complexity greatly: group-wide there would be just one architecture for on-board electronics, just one air-conditioning system, just one navigation system. Winterkorn wants to successively build each and every car in the Group on one of a maximum of four modular toolkit systems: the Up model lineup would be built on the subcompact variant, while models from the VW Polo to the Audi A3 and up to the VW Passat and Skoda Superb would be built on the MQB, and the Audi A4 through A7 or even the A8 would be built on the MLB. Sports cars are to be built on a sports car toolkit developed by Porsche (MSB), while luxury sedans such as the Porsche Panamera and Bentley models are to be based on another modular standard toolkit (MSB) also developed by Porsche.

Standardization of components installed across all of the Group's car brands will enhance VW's importance when it comes to engage innovative suppliers: "When you're talking these kinds of numbers, everyone wants to get a piece of the action," managers of leading automotive suppliers from Bosch to Schaeffler chime in unison to anyone listening.

Moreover, basic calculations show that Volkswagen absolutely needs to occupy new niches in order to increase sales and thereby fully utilize its workforce. Otherwise, productivity gains would lead to job losses, which is politically undesirable at Volkswagen.

The modular toolkit systems also offer environmental benefits: the cars produced on them should be lighter in weight and more fuel-efficient than previous models. Different versions enable finer regional diversification of the product lineup at different price levels. "With MQB, we will have the capability to select, in the world's regions and for each brand, whether a vehicle is to use a normal suspension or a high-end multilink version," Works Council chief Osterloh explained. In this way, the system permits a degree of qualitative differentiation that had been impossible before. Before the toolkits, to save on costs, brands had to resort to old platforms and components, with the Chinese-built Lavida for example being built on an earlier Golf platform, the PQ34.

Yet another advantage is that universal technical standards lead to uniform processes, stabilizing production conditions. Nonetheless, the modular toolkit architecture also harbors risks. The Toyota recalls that began in late 2009 certainly caught the attention of Winterkorn's quality managers – they are aware that these derailments are due to the fact that, to save costs, Toyota used as many uniform parts across its various model series as possible. A horror scenario for the German car maker would be a defect that slips into the modular toolkit systems, and is then reproduced in millions of vehicles; this would escalate damages exponentially to currently unimaginable dimensions. The risk is further increased by the large number of new parts that need to be developed.

To counteract this risk, Winterkorn is making quality a top management priority. At an employees' meeting at the beginning of February 2010 in Wolfsburg, he insisted on employees openly reporting defects. "We must take all quality problems very seriously and correct them as early as possible," the VW chief reminded his employees.

2.9 GROWING WITHOUT GETTING BIGGER

Volkswagen is faced with the same basic problem as Toyota: the majority of its employees and most of its automotive production capacity are located in its expensive but saturated home market. The Group itself expects Western Europe's car markets, apart from China its core sales region, to grow to 16.5 million units by 2018, from 15.0 in the crisis year of 2009. Eastern Europe, however, including Russia is expected to grow by an estimated 115 percent over the same period, to 6 million units. Volkswagen's experts expect Japan, Toyota's stronghold, to remain practically stagnant for another eight years, while North America (which for VW consists of the whole NAFTA region) is the only triad market that could compensate for some losses with an estimated growth rate of 55 percent to a volume of almost 20 million vehicles by 2018.

Accumulated global car sales, however, are slated to hit the 100 million mark by 2018 – thanks to dramatic growth rates in China, India, South America, and the ASEAN states. That is why new cars are increasingly introduced and sold in Beijing, New Delhi, or Moscow and less often in Tokyo, Paris, or Berlin; and despite efforts to globalize its operations, about half of the Group's total of 368,500 employees still work in Germany. In 2010, five out of six cars that the Group sold were in countries other than Germany, but four out of six were produced there. Intensive value creation in Germany adds to quality and image, but forces companies to export. The pressure to produce in the country in which the car is sold will intensify further in the future. "In 2018, the trend for employment will be that only 40 percent of staff will be employed in our home region, 60 percent abroad," said Head of Human Resources Horst Neumann about the jobs-shift to foreign countries.

There is still no permanent answer to the issue of volatile factory utilization in the saturated markets of Western Europe. Top management demands a 10 percent annual productivity gain, aiming to increase production without hiring new employees. Seven percentage points are to be achieved by increased efficiency, and the other three should come from reducing employment figures by staff turnover and early retirement. However, VW has to keep automotive manufacturing in Germany and Western Europe competitive, so significant productivity gains are necessary; and when it comes to building up new workforce, Volkswagen is targeting Eastern Europe, such as the Bratislava plant in Slovakia, for example. That is where the Group will be producing the new Up subcompact car models due to be launched in 2012 – along with the Touareg SUV and its cousin the Porsche Cayenne.

In the key markets of China, Russia, India, and the USA, demand can be met profitably only with local factories. Moreover, this will only be achieved if the majority of car components also come from the region. VW managers intend to increase the degree of localization to more than 80 percent, particularly in the USA: the Jetta had 70 percent localization at its production launch, while the new Passat should reach 80 percent according to Winterkorn. VW managers nevertheless admit that the local sourcing of parts desired cannot be attained in every region, this being especially difficult in Russia, where "the supplier industry and infrastructure must be significantly expanded," as VW sources put it. VW is aiming for 10 to 30 percent local sourcing in Russia by the year 2015. In its large Brazilian market VW intends to procure locally 80 to 90 percent of its parts, in China 60 to 90 percent, and 75 percent in India as early as 2012.

In order to avoid having to start from scratch each time a factory is built in a new market, Volkswagen is attempting to specify a basic factory design, as Toyota has done for decades with its standardized transplants (see Section 3.7). Typically, the basic VW plant has a production capacity of 150,000 cars annually, but output can generally be doubled by duplexing the factory design. After Kaluga in Russia (started in autumn 2007), Pune in India (started in early 2009), and Chattanooga in the USA (2011) the next standard sites to start production will be in China. One new plant after another far from the company's headquarters in Wolfsburg, and more and more new models from more and more brands – the tremendous rate of growth is putting the strength of VW's production engineers to the test. A number of model launches have been difficult, and Volkswagen urgently needs a whole slew of qualified production specialists, who are also in extremely high demand and being assiduously sought by other manufacturers. Furthermore, there is a lack of local managers in leading positions. "We must further internationalize our management team. We are getting to our limits here," confessed Winfried Vahland, Volkswagen's longstanding China chief, to the authors of this book.

In China alone, Volkswagen is doubling its production capacity to three million cars annually by 2014 at the latest. "We have not yet found sufficient numbers of qualified people in the country," said Vahland.

This makes uniform processes all the more important. The large factories in Ingolstadt or Shanghai produce so many different cars that there is precious little time between any two new vehicle launches, while previously there was typically a pause of six months. "That is why we need process standardization at all of our production sites and an experienced production launch organization, if we are to successfully complete all our new projects," former chief of production, Jochem Heizmann, explained. This applies to all vehicle classes. Just how an instrument panel is placed in the vehicle, how it is mounted on the A pillar and the sequence in which it is installed in the vehicle – that should be uniform for every vehicle, whether it is a VW Polo, an Audi A4, or a Skoda Superb.

A cultural chasm is opening up in the Volkswagen Group, a result of antagonism between the necessary corporate discipline and the creative space for individual brands. Technical standardization requires high volumes that can only be attained if everyone pulls together. However, the soon-to-be ten different brands of the VW Group cannot be managed centrally. Volkswagen controls the development and usage of the toolkit systems centrally, in order to keep costs firmly in hand. Branding strategies, however, such as sales-networks and model launches, are managed by each brand individually.

For his strong growth market, Volkswagen's China chief needs models that he can have assembled locally, and cost-effectively, out of a storehouse of components – as in the case of the Lavida. The two approaches must complement one another: standardization reduces the costs of tools, processes, and replacement parts, while localization reduces labor costs.

The history of the Audi brand in particular has been characterized by a guerrilla culture: the bestselling Audi 100, predecessor of the A6, was developed in contravention of orders from Wolfsburg. Group patriarch Piëch himself, during his years as Audi chief, pressed ahead with the introduction of lightweight construction methods in the early 1990s. Thus the A8 flagship got an aluminum body, although the head of Group Sales at the time, Werner Schmidt, opposed the idea. He thought that the new material could not be handled properly by customer service – on the other side stood Winterkorn, who was Chief of Quality Control at the time. Later, when Piëch moved up the ladder to Wolfsburg, Schmidt was one of the first managers to take his leave. The episode is typical of the very tense relationship between corporate headquarters in the German north and the ambitious subsidiary in the south. "Giving a brand sufficient space certainly only succeeds if one has learned the consequences of overly restrictive policies," said Winterkorn, who headed Audi from 2002 to 2007.

These experiences have consequences. Both he and Piëch had suffered tremendously under the requirements set by the corporate headquarters, the VW chief acknowledges: "Anyone who has experienced it personally never forgets this lesson." As Group leader, he wants to promote innovation without neglecting the corporate discipline needed for economic success. At the Frankfurt International Motor Show in September 2009, Audi presented an electric version of the R8 super sports car that it had developed, and it involved Wolfsburg directly, in contrast to earlier times. Winterkorn gives the brand a free hand, but he makes sure "that Audi does not take any other path in the area of electric mobility than that which we are taking as a group." And Piëch? From his position as chairman of the supervisory board of the VW Group, the patriarch still likes to stir up internal rivalry between the brands – and makes no bones about it. As far as he is concerned, to spur innovation and avoid complacency, Volkswagen and Audi are competing with one another, especially over diesel engines and dual-clutch transmissions.

2.10 INTELLIGENT BRAND STRATEGIES

A significant key to the success of VW Group lies in the management of its brands. No other corporation pursues such a consistent strategy of comparative independence of subsidiaries at the operational level. The balancing act consists of appearing to be different in the eyes of customers, while standardizing as much as possible technically behind the curtains. Today, the Group's combined purchasing volume lies at over US$100 billion (€70 billion), and when used properly, the VW purchasing department wields enormous power when it comes to dictating the prices and conditions in supply contracts.

Just how does the positioning of each of the individual group subsidiaries look? Winterkorn's philosophy for the core brand VW – which he, as Piëch once did in the 1990s, manages in addition to the overall Group – can be summed up like this: Volkswagen should combine German engineering craftsmanship with the Italian *grandezza* of highly acclaimed design chief Walter de'Silva, whom Winterkorn brought with him to Wolfsburg from Audi. Volkswagen wants to offer premium cars to the mass market.

Skoda and SEAT are there to serve the lower market segments. The Czech subsidiary Skoda's core USP is spaciousness at good value for money. Decades ago, Piëch set Volvo up as the standard by which it should measure itself. In an aside at the Geneva International Motor Show in March 2010, however, Winterkorn found fault with the excessive expense on interior fitting, for example, which was putting profits at risk. "We must ensure that Skoda remains profitable and holds us [more strongly] to the slogan 'Simply clever'", demanded the Group chief. New Skoda chief Winfried Vahland sees "spacious, functional and convenient cars for families" as the core value of the Skoda label. Within the Group's sales strategy, Skoda is to defend VW's position against fast-growing rival Hyundai/Kia. The image of SEAT, on the other hand, is weak and suffers from changing requirements: once a family car brand, then a sporty, entry-level variant, the brand is now slated to gradually receive a full lineup of models in an attempt to displace Renault, which has come to dominate SEAT's home, the Spanish market. Just how the company can recapture market share is illustrated by the mid-class model Exeo, launched in 2009, which is based on the predecessor of the Audi A4. The tools were taken apart in Ingolstadt, Germany, and reassembled in the SEAT plant in the Spanish city of Martorell.

SEAT is the black sheep of the Group, posting an operating loss in the first three quarters of 2010 of €218 million. Former Ford and Mazda manager James Muir wants to utilize upcoming model changes such as the Golf offshoot, the Leon, to give SEAT a stronger footing in the middle segments of the market and make the most of the capacities of the enormous main plant in Martorell. SEAT should turn a profit within five years, announced Muir, who has a reputation as a bit of a tough guy.

Audi is the Group's unquestioned profit maker, posting operating profits of just over €1.6 billion (US$2.3 billion) even in the crisis year of 2009, a result it then doubled in 2010 to €3.3 billion (US$4.7 billion); yet beyond this the Audi brand also represents Piëch's ambitions in the premium segment. He wants the subsidiary to overtake BMW and Daimler globally in upcoming years with technical innovations and a modern, dynamic, sporty image. In Europe Audi is ahead of its competitors, but in the USA, for example, Daimler and BMW have a much stronger market position.

Market areas above Audi are covered by the Group's luxury brands. Bentley represents the elegant, sporting, British grand touring and sedan culture, while Bugatti offers an exclusive racing car and French craftsmanship; Lamborghini presents itself as an Italian alternative to Ferrari, and Porsche is the epitome of the classic sports car, supplemented by the sporty four-seat Panamera and the Cayenne SUV. Nevertheless, the sudden drop in demand in 2009 shook the luxury segment to its foundations. Bentley posted an operating loss of €194 million, and it stayed in the red throughout 2010, running an operating loss of €245 million. Wolfgang Dürrheimer, former Porsche head of R&D, has taken over the helm at Bentley, and his first task will be to get the label back in the black and then join forces with Porsche's Panamera model series as far as technology is concerned.

Whether it really makes sense to have small, independently managed brands like Bentley and Bugatti is doubtful. A conceivable solution might be to reincorporate brand groupings as was done at Volkswagen under Winterkorn's predecessor Bernd Pischetsrieder. Porsche could then orchestrate the smaller, sports-oriented brands. At present, the free rein given to brand managers across VW Group is not inconsiderable. After the final release of a new model by the Product Strategy Commission, managers from Audi to Skoda take over and control the processes, from production to sales – with the exception of the modular toolkits they use for new models, which are intended to limit the degree of new component development in all brands, to save costs. Thus, the individual brands are spared a lot of R&D and testing, as long as they remain within the defined scope of the modular toolkit system.

Working segment by segment, Volkswagen is filling the niches in which the Group previously had no product on sale. After the compact Tiguan SUV, launched in 2008, in 2010 the Amarok pickup truck followed. Additional coupé and convertible variants of existing model series such as the Golf are planned, and with the Up an entire small-car family is in the product pipeline. The size of the Group enables high-image products such as the formidable Audi R8, based on a Lamborghini. Another example is the Volkswagen Concept BlueSport, a two-seat VW roadster with a centrally mounted engine, which was highly praised by the automotive press. Volkswagen presented it in January 2009 at the North American Auto Show in Detroit, and it could go into production in 2014. The BluesSport is proposed to take VW into a realm usually the preserve of Japanese manufacturers such as Toyota and Mazda, and show the world the latest iteration of VW's new, hard-edge design language, which was first showcased on the Scirocco coupé and then on the Golf VI. As Chairman of the Board of Management, Winterkorn concerns himself with clear positioning and an independent presence for the Group subsidiaries, e.g. at car shows.

Although the Group stands to have ten brands soon, the ambitions of Volkswagen's corporate patriarch Piëch seem to be never-ending. The Porsche heir is working on plans that go beyond this. "I am eying three other brands," Volkswagen's Supervisory Board chief has acknowledged, the truck maker MAN being one of them. Despite the regulatory trouble this tie-up will cause, when Piëch says he wants to do something, there is every chance he will end up being able to do it. So far, he has certainly never let anything knock him off his course. "I contemplate such ideas a very long time, sometimes thirty or forty years. Forty years ago, an engineer told me that a dual-clutch transmission could make a vehicle seven seconds faster with the same engine power on Nürnburgring," says Piëch. "For a long time, a succession of management boards failed to grasp the significance of this subject, until I was back in Wolfsburg from 1993 onwards."

On a number of occasions, Piëch has expressed praise for the motorcycle brand Ducati. He attests to the Italians' "incredible focus" that resembles that of sports car maker Porsche. VW has also been eyeing Alfa Romeo, with Fiat repeatedly saying that the brand isn't for sale. Winterkorn went on record as saying that Alfa Romeo is a "nice and interesting" company. Piëch has said that the German company is "monitoring" Fiat's plans for Alfa Romeo. His admiration for the brand is said to have begun early, during an apprenticeship with the automotive design company Italdesign of Turin – which has since become a 90 percent subsidiary of the VW empire.

2.11 PROCESS IS THE PROBLEM

As impressive as the formation of the multibrand empire Volkswagen may appear, and as focused and successful as its attack on Toyota may be in its execution thus far, the core brand VW is still lagging behind its potential. Former VW brand chief Wolfgang Bernhard, who is now a member of Daimler's board of management, once compared the VW brand – which in 2010 contributed roughly 4.5 million units to corporate-wide sales of 7.14 million cars – to a raw diamond that could be transformed into a shining jewel with the right polish. Yet after Bernhard's departure and that of ex-Porsche chief Wendelin Wiedeking – who was well known for causing disquiet among the Supervisory Board in Wolfsburg – no Group manager has dared to openly criticize the lack of profit orientation with regard to the core brand. This is clearly the downside of the departure of Wiedeking, who once aspired to take over the huge VW Group: his role as a necessary and substantially justified critic has been left vacant. More than once, this profit-fixated production specialist pointed to sore spots in the Wolfsburg empire, where, as insiders say, many projects continue to be "calculated more than optimistically." Minimum acceptable returns only exist on paper, while in practice these targets are often missed.

While the Group's best performing brands, Skoda and Audi, attained profit margins (EBIT) of 7 and 8 percent respectively in the pre-crisis year 2008, the VW brand did not even reach 4 percent. In the crisis year 2009, the figure was 0.9 percent: without China, the Volkswagen Passenger Cars brand barely scraped an operating profit of €561 million (US$808 million) from €65.4 billion (US$94.2 billion) in sales. Audi, on the other hand, secured over 5 percent return. In 2010, at least, the Group taken as a whole gained ground: operating profit was €7.1 billion (US$9.9 billion) with net revenues of €126.9 billion (US$177.6 billion). The group-wide operating profit margin was 5.6 percent after a meager 2 percent in the same period of the previous year.

Much of this improvement did in fact come from the core brand. The sixth version of the key Group model, the VW Golf, is now generating profits, with the previous model having generally brought mostly losses since its introduction in 2003. Works Council chief Osterloh says the new models being built in Wolfsburg, such as the Golf, are fully meeting their profit targets; this is also the case for the Polo being built in the Spanish city of Pamplona. "In the final analysis, whether I can build a car in Germany is really a question of returns. Today, we could actually build the Polo in Germany again," Osterloh said.

Certainly, Volkswagen's core brand did better even in 2009 than all other manufacturers of its size – with the exception of the Korean competitor Hyundai. Winterkorn's strategists no longer look just at the industry leader Toyota, for which a separate column is reserved in VW's sales statistics for better comparison, but increasingly keep an eye on Hyundai and its affiliate Kia, which attained a considerable rate of return of over 6 percent in 2009. The companies sold a record 5.74 million vehicles in 2010 and plan to release at least 12 new or revamped models in 2011. The Koreans' sales target for 2011 is set at 6.33 million vehicles – as much as the entire VW Group turned over in 2009.

In 2010, the VW passenger car brand made significant gains. As previously, the primary driving factor was China. Operating profits increased by more than €1.2 billion (US$1.7 billion) to €1.6 billion (US$2.3 billion). Net revenues increased by more than €11 billion (US$15.8 billion) to nearly €59 billion (US$85 billion).

Yet Wolfsburg has a habit of slackening discipline after progress has been made. An evaluation of business reports from 2000 to 2009 shows that in just three of ten fiscal years did the Group exceed the target rate of return on invested capital, which is set at the mark of 9 percent. In the other seven years, VW did not even earn enough to cover its capital costs, thus losing money. The negative balance from three good and seven poor years amounts to over €8 billion (US$11.5 billion). With this kind of previous form, a betting man would be unlikely to stake much on VW covering its capital costs in the current decade, even if 2010 and 2011 are good years.

Without question, the case of VW Group is substantially more complex than that of other manufacturers. No outsider really knows the modalities of internal accounting – including such knowledge as which research and development expenses the VW passenger car brand takes over in its own, divisional balance sheet, for the sake of its more profitable subsidiaries. In the premium sector, Audi in particular builds on R&D services provided by the Group, as well as on sharing components. Above all, in the high-volume segments of the compact class, the A3 and A1 Audi models benefit from their affiliation to VW bestsellers, such as the Golf and the Polo.

Exceptionally enough for a carmaker, VW continuously fine-tunes new models right up until their launch, for example the new Golf and Polo models. The fine-tuning is an executive approach dating back to Piëch's time as CEO. "This standard of perfection is not without its problems," concedes Winterkorn; competitors would call this practice a risky cost factor. Paradoxically, it is Winterkorn's improvements that repeatedly cause disturbances to processes at short notice. Heizmann points to the dilemma here: "We have a conflict. How much can we handle in terms of continuing model improvements without throwing us off course in terms of schedule and costs? This kind of optimizing right up to the deadline is a significant difference in comparison to Toyota."

Winterkorn's rationale for continuing to intervene is that when he notices an imperfection, he feels obliged to correct it immediately before production starts, especially since other imperfections are sure to go unnoticed. "When I see them, I suffer with the car for the rest of its life cycle," says the VW chief. However, late intervention is also a potential source of new imperfection that will then need to be corrected at great expense later on. The stability-oriented Japanese, on the other hand, must find such practices disconcerting. "A corporation like Toyota would never do such a thing. And the suppliers would not go along with it either. Everything is finalized a half-year before production launch. That is a very different approach," admits Winterkorn.

The particularities of VW's processes invoke the age-old question: which came first, the chicken or the egg? So long as Volkswagen keeps tinkering with its new cars right up to their production launch, process stability suffers; and because processes are unstable, managers find themselves needing to hastily react with hands-on intervention time after time. Whatever the answer, Piëch and Winterkorn have extremely high standards as engineers: this product fetishism is part of the Volkswagen myth, an inalienable part of the entire brand charisma. The company spent at least €1 billion (US$1.4 billion) developing the Phaeton luxury-class sedan, including its Transparent Factory in Dresden – and the rebates needed for sales and leasing of this model. Despite all this effort, the model remained a poor seller in Western Europe and the financial results simply do not add up. Even with a model facelift, costs will not be recouped within the foreseeable future. The Phaeton is priced to compete with an Audi A6, but is said to have incurred the production costs of an Audi A8.

Nevertheless, customers of such models as the heavy VW SUV, the Touareg, pay for the brand's claim to perfection. In the case of the Polo, this is hardly justified due to the brutal price wars in the small-car segment. The clear lesson learned from the unabashed success of VW's small, lifestyle-oriented Tiguan SUV, which former VW brand chief Bernhard groomed for profit right from its development phase, lies in offering the right product, with the right quality, at the right time under the umbrella of an attractive brand – and at a competitive price. Bernhard was right when he said time and again that technical excellence must not be allowed to be self-serving.

Winterkorn assesses the company's technical emphasis differently: "We are marked by absolute technical obsession. We want to make everything even better," is the motto of the VW CEO. And not only in production are product characteristics said to be very important to VW. Audi founder August Horch showed the way, quoted by Winterkorn as saying: "Whenever I've spent the day making a beautiful car, I reflect on how I can make it even better tomorrow."

Other explanations for the obsession with tinkering are less flattering. Volkswagen is said to simply lack Japanese virtues. "We lack the discipline for stringent systems," says a high-ranking VW manager who requested his identity not be revealed. One remedy might be the modular toolkit strategy, which has the potential to standardize and stabilize processes in the corporation. Volkswagen refuses to view the amount of refit work it requires in production – which is meticulously documented in a yellow folder for each car with a quality control card – as a deficiency of the production system, but rather as an expression of high quality standards.

The fact is that, especially in the USA, customer satisfaction comes in low. In an authoritative study by J.D. Power & Associates, Toyota's luxury brand Lexus was ranked first in the year 2009 before product recalls. The designated VW subsidiary Porsche took second place. VW ended up in 17th place, while the premium subsidiary Audi ranked a low 20th. A subsequent J.D. Power analysis in 2010 ranks Porsche in first place, Lexus in third, but Audi has dropped back to 26th place, and VW ranks a low 34th.

2.12 GOING THE "VOLKSWAGEN WAY"

The German car maker's production system, known as the "Volkswagen Weg" – which translates neatly into English as the "Volkswagen Way" – is intended to make production more efficient. The role model for this system was, of all things, Toyota's principle of continuous improvement, known as *Kaizen*. At Volkswagen, the system is called the "continuous improvement process" or CIP for short. In Wolfsburg, the Group spent €2 million (US$2.9 million) on building the 1700-square yard Lean Center, a model factory with a training ground attached. This is where employees from top management downward learn a lean production system, which they can then share with their colleagues across the globe. The construction of competitors' vehicles, especially those by Toyota – and increasingly also Hyundai of Korea – is scrutinized to identify scope for more efficient assembly. Even the administrative departments must "go the Volkswagen way" and optimize their organizational structures.

And that is only the beginning. VW plans to build Lean Centers for all brands and production sites. This means that the production team has to educate trainers and facilitators for the production system in a decentralized approach, to perfect the methods of the Volkswagen Way and promote it in multistage workshops – very much in the fashion of Toyota and its famous *Sensei*, thousands of which have been sent to overseas facilities and suppliers in order to guard the legendary Toyota Way (see Chapter 3).

The carmaker wants to offset the cost disadvantages of its high-wage production region in Germany and make it competitive through higher efficiency. Some domestic production facilities, toolmaking for example, are not believed to be especially cost-efficient, but there can be "no one else in the world who is faster and better in terms of quality than we are," VW manager Heizmann explained. Critics might still see in-house tool making as an expensive provision of work, contradicting VW's goal of becoming more efficient. Although Toyota is ahead of VW in efficient production methods, the former VW head of production is convinced that there are no soft spots on which Volkswagen is not already improving intensively.

A new performance parameter is intended to help avoid the traditional problems at the Group's most sensitive interface. "A crucial error that we have corrected was that our product development and our production did not communicate properly," says Winterkorn. Since 2006, the Group has been using the EHPV parameter to come to grips with this issue, the "engineering hours per vehicle" being the amount of time needed to build a car. The value of this parameter is intended to drop significantly with each new model, and on the current Golf VW managers achieved a 10 percent improvement; the modular toolkit systems should contribute a further 30 percent.

Of all places, it is in the largest factory – the main plant in Wolfsburg – that VW managers' hands are tied when it comes to modernization. The large production halls date back to the company's beginnings in the 1930s, and a modern car factory such as the one VW has set up in North America looks very different, being designed to implement state-of-the-art production methods. Wolfsburg's main issue is that progress needs to be rolled backwards, from the new plants such as Chattanooga, to the older ones in the home region. The obsolete structures at home, however, put managers in a quandary. "The employment situation at existing production sites plays a large role, and that can lead to a situation where we, under consideration of our productivity goals, react by increasing in-house production in order to protect jobs," Heizmann explained. Seats, for example, are produced by 1700 employees at the VW subsidiary Sitech in Wolfsburg, Emden, Hannover, Polkowice in Poland, and Shanghai in China. Axles come from the components plant in Braunschweig, transmissions from Kassel, engines from Salzgitter. "The sums just do not add up," says the CEO of a leading European automotive supplier. In confidential conversations, even VW managers cast doubt on the efficiency of the approach of producing standard components in-house and delivering them to the entire VW empire from Germany in order to protect over 150,000 domestic production jobs.

This criticism does not appear to have gone entirely ignored. At new production sites such as Chattanooga, the Group has moved away from in-house production in order to minimize investment, and at the opening of the Chattanooga plant, the man responsible for quality Marc Trahan was bullish: "Our suppliers are ultimately our responsibility," he said when asked about this uncharacteristic step. Yet even at its most modern facility, Volkswagen still insists on a comparatively high manufacturing depth. For the US plant, the company had considered outsourcing body parts such as doors – until now a taboo at Volkswagen – but the plan was rejected in the end, with in-house production being said to be more cost-effective. Nevertheless, the continuous improvement process based on the Toyota *kaizen* model has already streamlined some production methods: unnecessary machines have been eliminated, and employees do not have to work overhead any longer.

VW's Head of Production must perform a difficult balancing act. He is expected to stabilize production, but without being allowed to change its organizational structure substantially. Heizmann, the often impatient former Head of Production, quickly became a target for union representatives, which has consequences in the VW empire, as it is ruled in consultation with the workforce. In October 2010, Heizmann transferred his duties to Michael Macht, chief of Porsche until then. Michael Macht has specialized in lean production methods: as head of production at Porsche, he intensively studied Japanese automotive manufacturing practices.

The immense variety of car models and variants alone makes optimization of production a Herculean effort. A barcode at the front left of the body specifies the precise production characteristics, and the Golf can be ordered in over 3 million different versions – a level of variety that would be inconceivable for a carmaker such as Toyota. The general industry wisdom is that focusing on just a few amenities and special features makes production far more robust, entirely aside from the fact that complexity resulting from variety incurs enormous added costs, from development all the way down to sales. The braking forces of inertia are powerful in Wolfsburg, even though the Supervisory Board fundamentally supports the self-initiated Volkswagen Way, which is intended to standardize and disseminate production methods based on best practice examples within the Group. According to internal comparisons other production sites such as Braunschweig and Hannover, where car bodies are stamped and welded and light commercial vehicles are built, are further ahead.

Industry insiders note that progress made within the VW Group has been insufficient, and that the productivity gap between VW and the Japanese remains far bigger than the ever narrowing gap in sales. "We can't catch them," say even high-ranking VW managers. According to the industry Bible, the *Harbour Report*, when it comes to productivity, Wolfsburg continued to fare poorly in 2009, below the average for the automotive industry. While the VW Golf took a total of 32 hours' labor to assemble, according to figures from the *Harbour Report*, Toyota only requires about 22 hours to build its competing model, the Auris, in its European plants.

Volkswagen's former brand chief Bernhard held up Suzuki's Swift to his managers as an example of a nearly perfect small car adapted to different customer requirements. If he was hoping that this would be some kind of a revelation, he was disappointed. "VW has a terribly hard time building small cars that are profitable," Group managers admit. Even in the automotive entry-level range, Volkswagen's standards for features and technology mirror those in higher price segments. This is why the partnership with Suzuki is such a valuable opportunity.

2.13 SMALL COSTS BIG

Emerging countries such as China and India, with their fast-growing car markets, are not demanding just classy sedans such as the VW Phaeton. They also increasingly want inexpensive, entry-level models for just a few thousand dollars as an alternative to the usual electric bikes and motorcycles, although the Tata Nano (a few years ago the wakeup call for the industry to start building ultra-low-cost cars) did not sell well when introduced to the market. Nevertheless, other low-cost cars have, such as those produced by Dacia. The Romanian Renault subsidiary has surprised everyone with its success, especially in Europe, where its mix of attractive design and very low prices has been helping it eat into the customer base of mid-market manufacturers such as Fiat and, more worryingly for Volkswagen, SEAT and Skoda. Essentially, it has been Dacia's success that has forced the Group to see the potential in low-cost cars.

It is a big change for Wolfsburg: a Volkswagen for less than €5000 (US$7200)? Everyone at the headquarters used to agree it could never happen and would never happen. Even if the planned new Up small-family car were to appear in a slimmed-down version for emerging markets, VW would not be able to bring it in as low as €5000. "No fully fledged car could be produced to Volkswagen standards for a price point of less than €5000," said Works Council chief Osterloh when questioned on the topic in 2009. According to him, VW had a standard of quality and safety that the company simply could not let slip.

Yet this preconception has recently been challenged, and driven by Skoda's expansion and the desire in emerging markets for affordable cars, Volkswagen now needs to start pushing the development of low cost cars. In order to address the market, the Volkswagen engineers were forced to think low-tech, old-fashioned production technologies and work on the assumption of a high proportion of manual labor in local markets. Cars in this segment come with minimum safety standards, lacking expensive materials, and with a much lower level of passenger comfort; also they are built to be serviced far more often.

Even though these are old ideas on how to build cheap cars, they are seen as new at Volkswagen, which has spent the last 20 years moving away from them: and resistance to new ideas is a tradition at VW. Skoda-chief Vahland is the man behind the low-cost vehicle, and his way of re-thinking old vehicle platforms very much as it would be done at Toyota is the current main alternative to the vision of the centralized modular toolkit engine churning out cars for markets worldwide. The elegance of having both strategies at its disposal (i.e. modular toolkit design with homeland support and independent adoption of old platforms for low-cost, high-volume models) is that it will enable Volkswagen to expand its product line in markets like Brazil and Russia even further, all the while securing a profitable localized production base even without the world-class suppliers its toolkit models will need.

Within the brand family, the company with the biggest desire for growth in this segment will be Skoda, which has been told to reposition itself closer to the high-volume segment, distancing itself from the volume near-to-premium segment in which the core brand VW is operating. The current management team has extensive experience with low-cost markets, and this will also help utilize existing factories in Russia, where consumer purchasing power has not grown as quickly as forecasters had predicted.

In the emerging markets, it is crucial to keep technology in the vehicle to what is available locally. Ultra-light car construction techniques need to be forgotten; it is a case of back to the future, with the technical concepts looking very much like they did in the 80s in the West – heavy, and hand-built to a high degree. Building these vehicles in country is necessary for low-cost pricing in any case, since import tariffs still apply. In this segment, VW will be competing with OEMs like Hyundai, Dacia and, in Brazil, Fiat. The Chinese, too, will not wait long to enter this potentially lucrative business.

Parallel to Vahland's work on the new low-cost model, Group CEO Winterkorn commissioned an analysis of whether introducing a trimmed down version of the Up (which is still planned as a platform-built and not a toolkit vehicle), with local production in India, would pay off. Yet the new compact family is already a huge balancing act: as an elegant speedster of the Audi brand, a City version of the Up must appeal to affluent Western city dwellers, while trimmed down versions, presumably as SEAT and Skoda models, are meant to encourage urban mobility in emerging markets. Vahland meanwhile, with support of China CEO Neumann, is working with completely fresh material. In China, the new brand is set to target the segment below €8000 (US$11,000), being an important first step toward penetrating an area of the market from which VW is currently absent in China – and in most other emerging economies.

Vahland's cheap car will be built on a low-cost platform based on one or two old vehicles, a platform that can be localized to a high degree, and it will allow designers to experiment without interfering with the complex modular toolkit design system. In addition to this, the scheme will find a good use for old tools that have already served well and on which any further production will be highly profitable. Importantly, local engineers will be able to experiment with and adapt it, something the complexity of the new toolkits would make impossible on more up-scale models. Whilst the new design will probably have fewer airbags (and a consequently lower level of success in crash tests) and nothing more than simple air-conditioning, it will have basic on-board electronics. It should come in at way under US$10,000 and is likely to do very well in Russia, South America and, of course, China.

Volkswagen's historic weakness in the lower price segments is no coincidence. Since the Piëch days, engineers (so valued in the industry) have been trained to deliver high technical performance, and the Phaeton (handcrafted in Dresden) testifies to the technicians' quest for perfection. Winterkorn, who as VW Chief of R&D led the Phaeton project and then accelerated Audi's attack on BMW and Mercedes, values – like his supervisory board chairman Piëch – cars that radiate power.

Yet times are changing. The Group is trying to comply with customer desires in every region and in every market with its variety of brands and over 200 different models, trying all the time to cope with excess complexity and spiraling costs. The models in China like the new Bora are getting longer – because chauffeurs are common for customers buying in this class there – and yet, due to Group efficiencies and lower labor costs in the country, the vehicle costs little more than €11,000 (US$15,800). The sheer size of the Chinese market, and the strong regional differences, also mean that VW's range of vehicles on sale there is set to grow and grow: whilst the metropolises on the coast will be producing the same models as in Europe using the same toolkits (MLB, MQB); low-tech platform cars, like the 30-year-old Santana (still in production), will be shifted into the vast, comparatively backward hinterlands. This trickle down will see some odd VW pairings on Chinese roads, with extremely basic vehicles like the trimmed-down Santana alongside the very newest models, including electric cars.

In India, it is the Polo on which VW's hopes are riding. The Group is targeting Indian customers willing to spend more than €7000 (US$10,000), and production at the Indian Pune site was increased accordingly. Interestingly enough, the once-hyped ultra-low-cost Tata Nano has not turned out the success story it was expected to be. It has been beaten out by India's bestselling car, the Maruti Suzuki Alto, with a price tag of US$6200, and even the Hyundai i10 which is selling at US$7800. According to reports from India, Tata Nano sales in India dropped significantly in late 2010, due to various technical defects which have put off customers. So whilst it would seem that there is good potential in the Indian market for compact cars above the very lowest reaches of the price categories, it nevertheless remains to be seen to what extent Volkswagen can capitalize on that. The danger with the Up is that the concept is being stretched thin by the various demands placed upon it: will it survive being pulled by Audi at one end and by its role in the emerging markets on the other? If Volkswagen's recent history with low-cost projects is anything to go by, it will be pulled too far in Audi's direction to do well in the more price-conscious, less luxury-orientated mass-market segments of the emerging economies. The days in which Volkswagen thrived on the world car the Beetle are long past.

Indeed, with VW's top management becoming conscious of just how difficult it will be to integrate the micro segment into its culture and still sell at a profit, in December 2009 the Group entered into a partnership with the very suitable Suzuki Motor Corporation. Suzuki seemed to offer VW the possibility to change its mentality. Almost every second car sold in India comes from Suzuki's subsidiary there, Maruti Suzuki – with this in mind, it is not at all surprising that Piëch waxed lyrical about the brand he considered so technologically brilliant for so long in the lead up to the deal. The stake in Suzuki was meant to open the door to new possibilities in the growth markets for VW, all the more since the Japanese corporation has considerable competence in the field of motorcycles and K-cars. The latter are microcars – about the size of Daimler's Smart – which, until now, have only really been popular in Japan. They offer optimum usability in a very limited space, and "this is something that could be important for Volkswagen," Winterkorn explained before the deal in late 2009.

VW's CEO was thinking intensively about good-value small cars, yet not an alternative to the Tata Nano. "Volkswagen will never produce a cheap car of this kind – we must not and will not go down to the Nano-level, not even in cooperation with another manufacturer," he stressed. It would simply not fit with VW and its standards, which include minimum safety standards amongst others, and these "cannot be offered by cheap cars. I don't think even Suzuki can do it."

Volkswagen executives had hoped that the tie-up with Suzuki would be a strategic step that would allow them to resolve this dilemma, to learn how to design the kind low-cost car the Group needs to offer in emerging countries, yet without compromising on standards. Yet in fact, hopes about swift and effective cooperation on a technological level have given way to disappointment: there are no concrete results and none likely until the medium term at the very earliest. The Suzuki deal has turned out to be little more than a purchase of shares – with a lot of warm words about vague ideas of working together thrown in for free. Not that anyone in Wolfsburg is complaining about financial aspect of the 19.9 percent stake in Suzuki, of course, with the carmaker going from strength to strength and continuing to be highly profitable.

In fact, according to VW sources, top management may well be prepared to swallow up Suzuki completely should the 80-year-old Osamu Suzuki, current boss, decide to retire. This would add 2 million units a year to VW sales, and the Group would have a production branch in Japan, providing considerable further potential for synergies in terms of components and modules. Until this happens, however, VW must show that it can organize cross-continental cooperation to the benefit of both companies: this is something that, even after decades of running various brands with very different management cultures, VW still needs to prove. VW also looks back at discouraging experiences with joint projects, especially such mishaps as the recall of US VW Routan models built by Chrysler.

According to managers in Wolfsburg, the first months of cooperation with Suzuki in 2010 already threw up a slew of cultural issues, which are undoubtedly one of the principle reasons behind the stalled exchange of technology. The Japanese corporation is said to have demanded extensive access to Volkswagen technology without revealing any of its own technological achievements, and VW's longest-serving manager Detlef Wittig spent much of his time in Suzuki management meetings mediating between the two mentalities. Meanwhile, a liaison office in Wolfsburg was established to bring ideas together, and at the Geneva Motor Show early in March 2010 the Japanese patriarch Suzuki had his photo taken with Piëch and Winterkorn in a public show of unity. In order to show how close these men at the top are, Piëch took the limelight at the Geneva Motor Show 2010 and praised Suzuki for putting top-of-the-range technology into the cheap-car segment whilst other manufacturers were losing money with small cars. "We're all just trying to survive as best we can," admitted the boss of the Japanese carmaker, "and I thought to myself: with times as difficult as they are, I knew we needed a partner." Volkswagen is not just a partner, according to him, but a great teacher, too. Warm words, but words which look somewhat hollow a year later and have brought about neither the transfer of technology Suzuki was hoping for nor the change in mentality in the low-cost segment that Volkswagen wanted to initiate.

Meanwhile, driven by the necessary expansion in emerging markets, and without any input from Suzuki, VW is going ahead with its own low-cost car line-up for BRIC (especially China). If the new Chinese brand really comes to life, this new, low-cost car would fit into the strategy. Or it could turn out to be neither fish nor fowl: too expensive to be a real low-cost car and yet too cheap to be anything else. Whatever the case, VW has finally realized that producing a cheap car is a feasible business option.

2.14 LUXURY BECOMES OBLIGATORY

Until now, Volkswagen has always managed to launch the innovations the markets asked for, even when this has taken considerable effort. Time and again, initial clueless inertia gave way to highly focused action: not only has Volkswagen managed to introduce new technologies like four-wheel drive, lightweight construction, or low-consumption motors with direct injection (known as TFSI) and direct shift gearboxes (DSG) with seven speeds, but VW models have become symbols for entire generations of drivers – above all, the Beetle, the microbus and the Golf have started a cult and attracted fans all over the globe.

The contrast to Toyota and its traditional philosophy here could not be stronger. When the going gets tough, the German group is willing to take risks; the Japanese company, meanwhile, finds it difficult to change a long-term strategy, even when under public pressure as with the quality issues in 2009 and 2010. Endurance is often key in the automotive industry. In the 1970s and 1980s, the Golf, Polo, and Passat were competing with rivals from GM's European branch Opel/Vauxhall, Ford, and Fiat; yet at the beginning of the 1990s, Piëch quite consciously took Volkswagen up a level compared to its competitors and told his managers that they were now to start comparing themselves with Mercedes-Benz. This paid off. As Mercedes has started to lose strength in recent years, VW is seen as much improved, with a Golf today offering better interior quality than a Mercedes A or B Class. So it is not surprising that Mercedes only sold 222,400 units of its two compact models in 2010 (compared to 250,000 in 2008), while Volkswagen sold about 1.6 million Golfs, Jettas, and Boras in 2010.

The premium VW takes for its move up the scale can be accurately expressed as a price on the windshield that is 5–10 percent higher than other mass manufacturers like Ford; and VW needs this premium if its complicated processes are to remain profitable. Group CEO Winterkorn has accepted that this means there will be some crossover between VW and Audi, yet he wishes to avoid cannibalization of the kind seen between VW and Skoda when both were switched by Piëch to a platform strategy. Winterkorn is relying on the internal, high-level, Product Strategy Commission, comprising all brand heads, to prevent cannibalization of brands within the Group's realm and adjudicate their individual claims at an early stage. It is a tricky task. A certain level of crossover is good for the Group's image, but competition between brands can cost money. Every two years, price positioning between VW and Skoda is examined, with the new Skoda chief Vahland having the task of taking the Czech manufacturer back to its core values. This may also go some way to explaining the Group's rethink in terms of the low-cost segment, driven as it has been by Vahland.

Apart from this, the Group's ambition to be the leading manufacturer in terms of innovative automotive technology is clearly advantageous. If management is defined as the art of orchestrating a high number of people and focusing them on one common goal, Volkswagen has a very successful corporate management. Furthermore, the clear drive behind the manufacturer saves arguments with suppliers and focuses the strength of the Group.

Integrating Porsche into VW will require no revelation about premium cars for the VW Group. Volkswagen has already been there and done that: in late summer 2005, the Group brought the Bugatti Veyron to the roads. Constructed in a new factory at Molsheim in Eastern France, the Veyron's 1001-horsepower engine showed the ambition behind the car: it is intended to be the fastest mass-produced car on the planet. In 2010, the offshoot Bugatti Veyron Super Sport, equipped with 1200-horsepower, reached a speed of 268 mp/h (431 km/h) on Volkswagen's test track – a world record.

2.15 THE SEARCH FOR THE RIGHT VW ENGINE

Volkswagen's marketing strategists do not like scrimping on showy effects and are known all over the industry for their love of glamour. On February 10, 2010, German television legend Thomas Gottschalk guided 600 hand-picked guests through a premiere evening of a very special kind for the launch of the newest version of the Touareg SUV. Nevertheless, as VW executives were concerned, Germany's most popular TV host, along with star violinist David Garrett, were probably just about good enough for the presentation of the first hybrid version of a VW mass-market car: Volkswagen's love of lavish launches means that the Group's PR department would, even if Elvis were to be found alive, well and willing to perform, still be shopping around for someone even bigger. Behind the scenes at a fancy Munich hotel, the car experts were nevertheless rather excited: this was a very late first hybrid model. With the Prius, Toyota had been offering one for 13 years and the German group's technological achievement is not exactly spectacular; the electric motor for the Touareg comes from the German-based supplier ZF, and the batteries are cookie-cutter nickel-metal hydrides (Figure 2.3).

Nevertheless, despite this kind of delay with headline green technology, the Group has never had reason to be ashamed of its CO_2 emissions, despite its being so late with hybrid technology. Of all the global car firms, only the Korean manufacturers Hyundai/Kia, who have few luxury cars in their lineup,

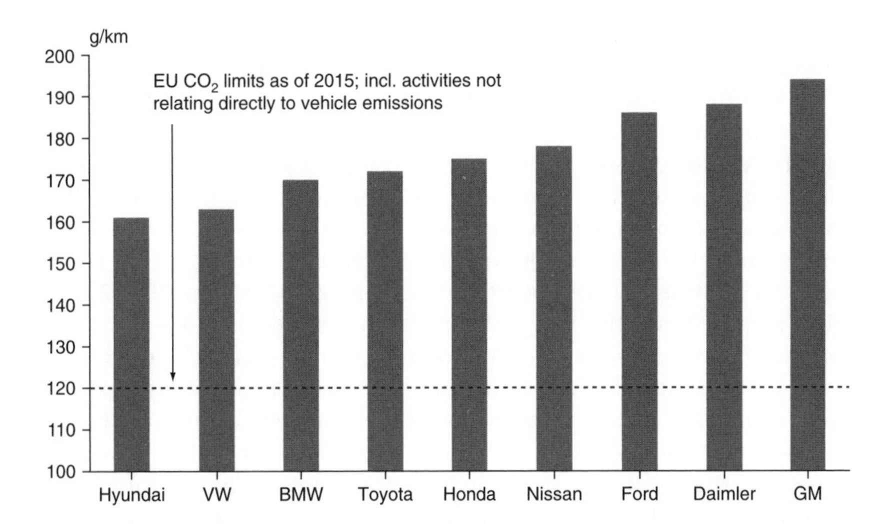

FIGURE 2.3 **Estimates of average carbon emissions of fleets from various European and American manufacturers (no standardization with respect to variations in road tests, 2008 data).**

Source: European Federation of Transport and Environment, ACEA, and US Department of Transport, analysis by PA Consulting.

do better than the carmakers from Northern Germany. VW has been very good at turning standard gasoline and diesel engines into TSIs, TFSIs, and TDIs.

The importance of reduced fuel consumption is something that Piëch recognized relatively early, driving forward the development of the VW Lupo, which went into mass production in July 1999 as the Lupo 3 TDI – Volkswagen's first "three-liter car" (more than 75 mpg). He also pressed ahead with the Audi A2, yet both of these pioneering vehicles were expensive single projects that flopped commercially. Customers were not ready to pay a (relatively high) premium for the technology. The gap in the Volkswagen product range has to be filled anew, and VW is placing big hopes on the Up new small car family. The direct successor to the unfortunate technology platform of the VW Lupo enters production in late 2011 in Europe. VW plans to sell 500,000 units of the Up family in one full year of production.

For its special line up of low-consumption vehicles the Wolfsburg-based manufacturer has coined his own term, "Blue Motion," and this has been accepted by the market; the advertising team even had a song composed: "Blue is the new green." Blue Motion models are made by combining an entire spectrum of technological improvements such as optimized motor management and driveshafts, start–stop systems, longer gear transfers, and better aerodynamics thanks to undercarriage and rear-axle casing as well as lower resistance on the ground. To keep this profitable, VW sells Blue Motion models in Europe for a few hundred Euros more, and the customer pays not just for reduced fuel consumption, but for the ecological image. In order to get more performance with less engine displacement, the engineers have been increasing engine pressure to new heights year after year, resulting in snappy little motors that are highly fuel-efficient. Combined with innovative DSGs, this allows Volkswagen to offer impressive savings in consumption and emissions: in 2010, 179 of VW's 200 models emitted less than 130g CO_2 per kilometer, 93 of those 179 even emitting less than 120g CO_2 /km (193.12 g/mi), and 16 of these below the 100g CO_2 /km landmark (160.93 g/mi).

Meanwhile, since Piëch's time at Audi, Group engineers have been looking at ways to reduce vehicle weight, since new technologies and safety legislation have continuously added extra pounds at the other end. After Audi became the first carmaker to produce a car with an aluminum body with the Audi 100 prototype in the mid-1980s, the Group has been using not just aluminum, but also magnesium, carbon fiber, and light construction steel in its components; it has even been experimenting with forms of metal foam. In March 2011, Volkswagen took a stake of 8.1 percent in Germany-based SGL Carbon, which produces carbon fiber for car chassis. The €140m move is seen as striking given that two years ago SGL formed a joint venture with VW's rival BMW. The move highlights how lightweight materials are becoming a central part of the battle for automotive technology. According to Piëch, VW took the stake in order to gain deeper access to lightweight materials. While carbon fiber is still mainly used in the aerospace industry and for high-end sports and racing cars, it is moving down to mass-market, premium car producers, largely because of the need to save weight in electric and hybrid cars. Conversely, Audi recently announced a cooperation deal with German engineering conglomerate Voith to build lightweight materials. Voith also owns a 5 per cent stake in SGL.

Along with the rest of the European automotive industry, however, Volkswagen underestimated the trend toward hybrid vehicles, despite the fact that the Wolfsburg engineers were pioneers in this technology 20 years back and, in 1994, brought the first hybrid vehicle, the Audi 80 duo, onto the market long before Toyota. Yet the car was so expensive that it was practically impossible to sell, a problem that did not disappear with the Audi A4 duo in 1997, costing as it did DM60,000 (€30,000 or almost US$43,200 in today's money). The customers were not convinced by the new technology, and only 90 of the cars were sold. The poor sales convinced Volkswagen that there was no market for hybrid vehicles and the Group then concentrated on optimizing the gasoline engine and developing diesel direct-injection technology.

In January 2011, Volkswagen chose the Qatar Motor Show to unveil its XL1 Super-Efficient Vehicle (SEV), the third evolution of its one-liter car strategy, some version of which the company will produce in two years' time. The XL1 is a small, two-seat vehicle with a diesel-electric hybrid powertrain: a plug-in, parallel hybrid, with an electric motor and batteries that allow the car to travel up to 35 km on electricity alone (although the motor is mainly there to boost the two-cylinder diesel engine, rated at just 47 hp). According to VW, it has breathtaking fuel economy: merely 1L/100 km. Both Winterkorn and Piëch confirmed that the car will be produced in limited numbers, first for the German market. To keep the weight down, VW used a carbon fiber-reinforced polymer *monocoque* with a production process it says makes the car far cheaper to produce in higher numbers. The XL1 weighs a scant 795 kg, and the streamlined body has a drag coefficient of just 0.186, compared with the Golf's 0.31, for greater fuel efficiency.

Analysis of the profitability of hybrid and electric vehicles

Figure 2.4 shows annual costs for mobility (based on the use of an average car) for an end-user. It compares the various costs for vehicles followed by the energy costs (fuel or electricity) per year, assuming that the consumer purchases one vehicle in the given year.

The decisive question is: at what point is it worth the extra cost involved in purchasing a hybrid vehicle in terms of fuel saved?

The results are surprising. Since gasoline in the USA is subject to far lower taxes than in Europe (see scenario 1), customers buying a hybrid will not benefit from major savings before 2030. A full tank of gas costs far less in the USA,

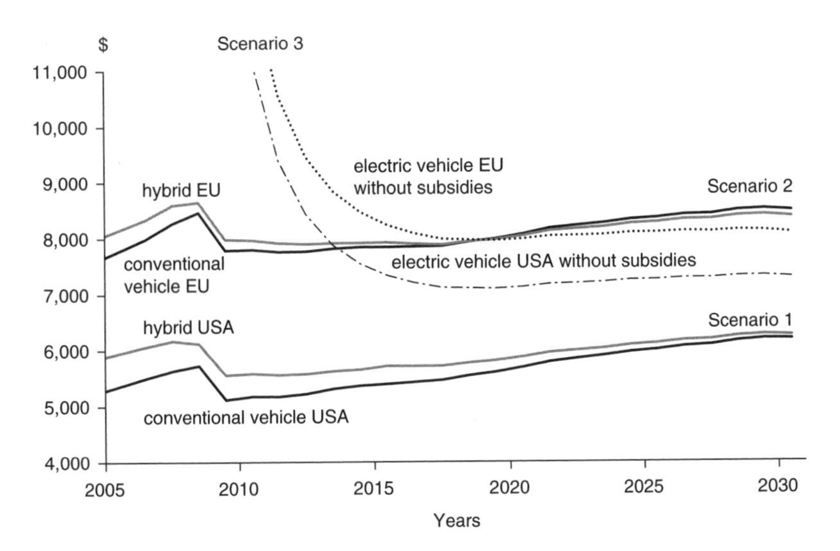

FIGURE 2.4 **Total cost of ownership scenarios for conventional, hybrid, and electric vehicles in USA and Europe.**

and hybrids simply do not save enough fuel. In Europe, things are different (see scenario 2): a hybrid will be profitable there from roughly 2017 onwards (this takes no account of penalties imposed by the European Commission for not keeping within CO_2 limits). Scenario 3 shows the costs involved in purchasing and operating a fully electric vehicle with a range of about 150 km (93.21 mi), not taking into account the current subsidies, which might expire in the coming few years. This kind of vehicle should become noticeably better value in Europe by 2018 and, if it is subsidized, the point at which customers consider it worthwhile to buy an electric vehicle could even move forward to 2016.

Figure 2.4 above shows annual costs for vehicles as paid for by the consumer. These figures assume an annuity for the purchase of the vehicle itself, and for average gasoline (super unleaded) or electricity use. For the cost of gasoline, we used the *Annual Energy Outlook* 2010 predicted oil prices given in the Early Release Overview (Reference Case) in December 2009[i]. German electricity prices used here are based on a similar study and show relatively constant electricity bills for consumers between 2010 and 2030 adjusted for inflation; we recalculated using a standard inflation rate of 1.5 percent over the 20-year period. The same was done for US electricity prices. We did not take into account any costs relating to maintenance, insurance, road tolls or other expenses.

Annuities were based on a constant interest rate of 8 percent, spread over an (optimistic) vehicle and battery life of 10 years, with no battery renewal cost added for the electric vehicle.

The prices for new cars in Europe correspond to information in the DAT 2009 Report; for the USA, we used the NADA Data book 2009. The average yearly distance traveled (15,170 km /12,000 miles) is based on figures from the DAT Report 2009, Autohaus Extra (Germany) and information from the EPA in the USA.

The observations made for hybrids in the EU and the USA juxtapose fuel costs for conventional vehicles and hybrids in the USA (observation scenario 1) and Europe (observation scenario 2).

In order to produce a standard figure, fuel consumption was modified by correcting factors according to the market in question: we took real consumption into account, as well as correcting for the fact that the EPA driving-cycle figures are far more realistic than those produced by the NEFZ in Europe. We assumed that fuel taxes will remain largely constant in both the USA and the EU countries, which produces a consistently lower gasoline price in the USA than in the EU. The hybrid vehicle figures taken for the study were a mixture of the Prius, the Insight, and other models in development known to PA Consulting Group. The changes in costs for the engine and the battery in the hybrid were calculated differently.

The assumed gain in efficiency from hybrid vehicles climbs step-by-step from 20 percent at present to 27 percent in 2015, where it then remains. The improvements come from optimizing supplementary consumption (i.e. climate control, cooling) as well as a higher degree of engine efficiency and better battery technology. PA assumes that hybrids, too, will be using lithium-ion batteries instead of nickel-metal hydrides by 2015, when lithium-ion battery cells will be cheaper per KWh than hydrides. The efficiency advantage of hybrids can be expressed as a cost advantage in the calculations for gasoline consumption, but must be offset against higher purchase prices. Furthermore, none of this takes into account penalty taxes (such as those decided upon by the EU from 2015 onwards), nor subsidies offered by governments to increase the uptake of hybrid vehicles.

Assuming constant price decreases for batteries and hybrid technology, a hybrid system will be cheaper than a conventional vehicle by 2020; in the USA, the lower gasoline prices will mean that no break-even point is reached for hybrids within the timescale covered by this study.

A calculation for an electric battery car (observation 3) was added based on the following data: Capacity of 25 KWh; battery cost developments as shown in Figure 2.4; a range of 5 km (3.1 mi) per KWh. The range is much lower than the standard 7 km (4.4 mi) per KWh assumed in much of the literature about electric cars, but we think it is closer to the range achieved in real conditions using cooling and climate control. This range factor is increased by no more than 1.5 percent per year, as a number of recent practical tests show.

The cost of the electric car without battery (€12,500 or US$18,000, including all applicable taxes) was calculated with an additional €2500 (US$3600) per vehicle for development as well as a supplementary €1500 (US$2160) to pay for infrastructure such as chargers. The battery costs are derived from multiplying the assumed capacity with costs per KWh in accordance with the battery cost curve. Costs for a replacement battery at today's technical standards were not taken into account. Furthermore, we did not include the assumption that manufacturers will react to falling battery costs by increasing range.

The modeling shows that electric vehicles can be in a position to offer savings compared to both conventional and hybrid vehicles as early as 2018 within the EU. This reflects only costs and does not show the disadvantages many drivers may subjectively feel in terms of reduced range and lengthy battery-charging times.

If electric vehicles are heavily subsidized, the break-even point could even come earlier.

In addition to the difficult question of whether or not an all-electric or hybrid car will ever pay off for the customer, there is a specific problem with gas-electrics: the hybrid effect – which involves saving energy while braking, and reusing it through the additional electric motor when accelerating – only works when the driver spends a lot of time braking. This should occur relatively rarely on longer highway journeys in Europe, for example; but in the continuous stop-and-go traffic of Asian megacities, hybrid technology has a clear advantage.

For the Wolfsburg carmaker, at any rate, hybrid cars seemed unprofitable and were not taken further after the early hybrid adventures of the 1990s. The way forward seemed to lie in light vehicles and optimizing gas and diesel engines. Toyota, however, showed a lot more endurance and even managed to turn hybrid technology into an indisputable competitive advantage. What has become clear is that customer expectations, lawmakers, and overseas markets are forcing VW to catch up with Toyota at a breakneck pace. Whatever Volkswagen may think about the poor commercial prospects of electric vehicles at the moment, it has realized the need to have them in its line-up and be ready to produce more in a short production cycle. What would happen if several European – and perhaps even Chinese – cities suddenly legislated against solely gasoline-driven vehicles? Before, this was the kind of nightmare scenario that woke automotive engineers in the middle of the night before they realized it was all a dream and went back to sleep. Following congestion charge legislation in major cities like London and Singapore, however, and approving noises from a host of others, the idea of banning solely gasoline-based vehicles from city centers is slowly drawing closer and closer to the statute books. In other words, only consistent electrification will help the Group keep up in the future, and this means that VW has really had to rethink its approach in the space of just a couple of years.

In order to lay a path toward the electric age, the Wolfsburg manufacturer has created a new cross-group task-force, partly because Winterkorn was not happy with the slow rate of progress. From December 2009 to summer 2010, the former Continental CEO Karl-Thomas Neumann was in charge of VW's electric car development and mapped out a strategy for taking Volkswagen into this new era of green cars. Then Neumann became chief of VW's Chinese operations, and Rudolf Krebs, who had been head of VW's central engines facility near the Wolfsburg headquarters took over. Across the Group, its hierarchies, and its brands, he is now responsible for making sure that VW catches up with the trend toward battery-driven cars.

Volkswagen has to tackle four main strategic challenges:

- affordable engine hybridization
- clear-cut product strategy for electric vehicles

- future-proof battery sourcing, and
- a new mobility laboratory to test new business models.

Affordable hybridization

On the journey toward an electric car, there is currently no route around the combination of gas and electric engines. Legally binding CO_2 limits that the EU is introducing, which will severely penalize non-compliant manufacturers from 2015 onwards, make the leap to next-generation cars even more necessary. Governments all over the globe, from the USA to China, regularly review their environmental policies, and CO_2 limits are increasingly on the agenda. This means that VW needs a stringent strategy for hybridizing its entire range, and the first Group product to be equipped with a combined, gas-electric powertrain was the Touareg. At the Geneva Motor Show 2011, Audi introduced its new, hybrid, mid-sized SUV Q5. Winterkorn's job is to find affordable ways to finally hybridize each Group product.

It can reasonably be expected that the new VW entry-level Up model will come with a mild form of hybridization (the start–stop system), but that in the compact class of the VW Golf (using the modular transverse toolkit), the new model will be fully hybrid, like the coupé shown in Detroit in early 2010; it has to compete with the Toyota Prius and Auris, after all. As for large executive cars, the Group needs plug-in solutions for its top-of-the-range models like the A8 and the VW Phaeton. Plug-in vehicles, charged on the electricity grid, allow cars to run on electricity for the first 15–30 miles of a journey, reducing fuel consumption. Plug-in systems are supposed to turn even large and heavy luxury vehicles, such as the new Mercedes S 500, into comparatively environmentally friendly vehicles.

"Volkswagen is going to be the automobile manufacturer that takes hybrid technology out of its niche," said Winterkorn, taking a potshot at Japanese rivals such as Honda or Toyota, whose hybrid flagship Prius has nevertheless written automotive history. Aiming to make a virtue out of necessity, he went on record with no small amount of bravura: "It's typical Volkswagen. We're late into the market, but when we're there, we clean up!" he boasted in 2010.

Nevertheless, especially in the compact class, Wolfsburg has something of a dilemma on its hands. The Group desperately needs a good hybrid offering in order to match Toyota, but the budget is tight. The German manufacturer is simply not prepared to spend a lot of money for a technology that has uncertain economic perspectives. Furthermore, the basic Golf is already pricier than the Toyota Auris in its standard design, simply due to its higher production costs, and this does not make it any easier to spend a lot of money on fuel-saving technology, especially since the price that car buyers will be willing to pay for it remains unclear. VW has already had the painful experience of being ahead of its time with projects like the three-liter Lupo, which swallowed enormous amounts of money without yielding the expected market breakthrough for green-ish vehicles at the time.

The effects on costs and fuel economy of hybridizing the Golf and the Auris, the two most important vehicles under discussion, are shown in Figure 2.5.

A compact car based on Volkswagen's modular transverse toolkit, such as a seventh-generation Golf, in a hybrid version, will cost almost €3000 (roughly US$4320) more to manufacture than a Golf with a single, gas-driven engine, resulting in a price that is around €5000 (approx. US$7200) higher for the customer – after sales costs, warranty, logistics, marketing, and a profit margin have been added; this is roughly equivalent to the current pricing of the Toyota Prius hybrid. Nevertheless, thanks to Toyota engineers' lead in

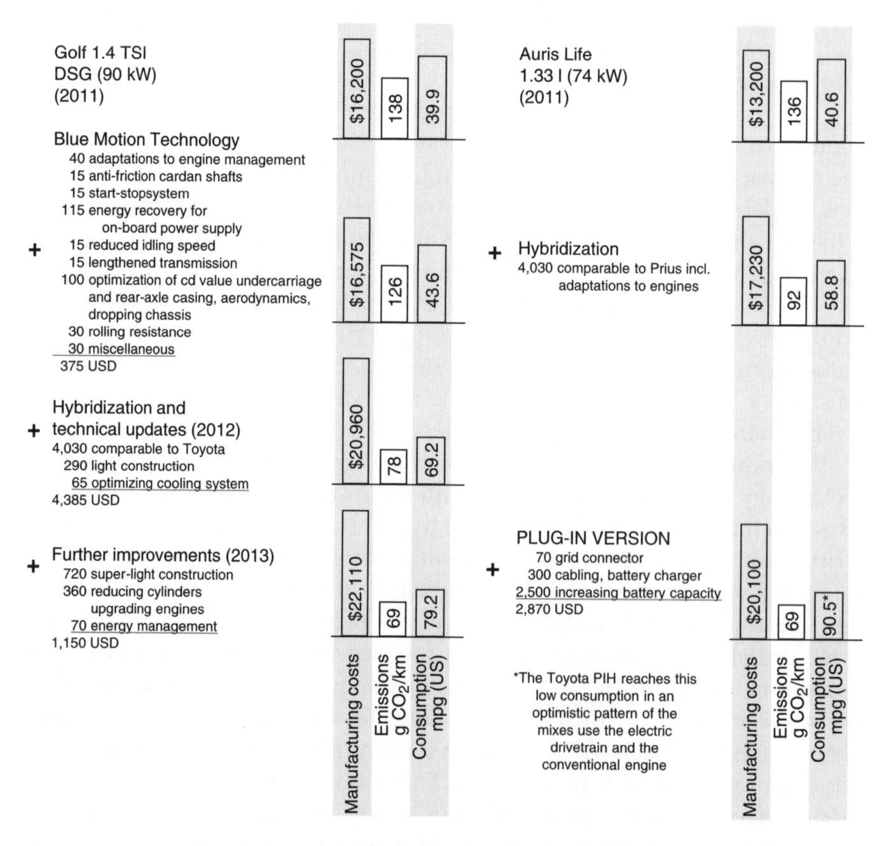

FIGURE 2.5 **Calculating the costs of lowering CO_2 emissions through 2015.**
Source: PA Consulting Group (all values are estimates by PA Consulting Group).

hybridization experience, the Japanese hybrid model will offer a noticeably higher level of performance in the same size category. While, according to industry sources, Volkswagen's electric engine will reach roughly 20 kilowatts, Toyota is already hitting 60 with the Prius III. Although the engineers in Wolfsburg are a step ahead of the smaller Japanese rival on this field, Honda, with its mild hybrid Insight, Toyota is out of reach.

The overall development of costs is worth noting, because it is not just the hybrid powertrain system that will make the compact car more expensive, but all the other measures taken to push down fuel consumption. Yet despite combining all these other technologies, VW is still unlikely to comply with European Union CO_2 legislation; the number of larger vehicles, four-wheel-drives, and premium models in the fleet is simply too high. This means that, in the medium term, there is simply no way around hybridization. VW's scope of action is, however, limited inasmuch as sales prognoses for hybrid vehicles are not rosy, with sales chief Klingler's staff estimating demand up to 2015 at no higher than a five-figure sum annually. In this scenario, hybridization is a serious financial risk.

The prospects for full-size cars are better. Social pressure will work for VW here, forcing customers to demand hybridization of these prestigious status symbols. With the base price already at a much higher level, additional costs are easier to pass on to the customer, and car owners in this class appreciate innovative technology. At the Geneva Motor Show 2010, for example, Audi presented a concept hybrid A8. Nevertheless, its fuel efficiency and reduced emissions came at a very high price: the plug-in approach for top-of-the-range models like the A8 and the VW Phaeton will cost roughly €10,000 (US$14,400) extra in the coming years.

A clear product strategy for electric vehicles

As indispensable as hybrid engines will be as a bridge technology, hybrid is just that: a bridge to the electric age, and not the destination itself; and while building this bridge, carmakers must also concentrate on developing all-electric cars. Here, VW has decided against constructing a flagship, as hybrid pioneer Toyota did with the Prius, and will instead roll out electric powertrain variants through its modular toolkits, producing electric versions of existing models. The Group is following the tried-and-tested method of creating a lot of different models out of the same standard components. Furthermore, this will spread development costs for battery-driven cars across the Group's brands.

Nevertheless, a flagship electric model would come with the advantage of being entirely conceived within the confines of electric propulsion, thus using the strengths of this technology to full effect instead of compromising with existing combustion engine requirements. Furthermore, this kind of project would most likely result in a boost to the Group's public image. But once bitten, twice shy; Winterkorn was close to the action as the top-of-the-range Phaeton flopped; there is no guarantee of success for one-off models in new automotive categories. Nevertheless, VW top management has discussed the issue of strategy for going electric, and has shown preference for the strategy of developing electric models out of existing ones within the scope of its modular toolkit, instead of designing a new, singular, flagship car.

A brake on the electric strategy could well be internal resistance at Wolfsburg. The proud brand bosses like the new Skoda chief Vahland or Audi boss Stadler have got their eye on the new key technology; the Works Council is concerned about keeping the workforce in German component factories occupied, and Volkswagen's tacit self-obligation to maintain employment forces the Group to a high degree of in-house production. A substantial part of traditional value creation is based on combustion engines and conventional transmissions produced in-house, but with electric cars, these two pillars would collapse – there would be nothing left to support the 10,000 to 20,000 employees who produce components for gas engines. The Group would not be able to integrate this workforce into the heavily automated production of modern batteries and electric motors. This discussion nevertheless leaves out the key strategic question of whether carmakers should allow these crucial components to be produced elsewhere. Works Council chief Osterloh stated that "sooner or later there will have to be a discussion about whether we finish batteries ourselves or simply buy use-ready products."

Originally, the idea was for the new compact Up family to be the first available in an electric version, but there was resistance among the members of the supervisory board. Faced with the financial and technical efforts required to bring this model to mass production, VW would have to concentrate its strength, the board stressed; instead of electrifying the base model of the Up, the Board demanded an electric version of the MQB. According to the Board, as the basis for high-volume models all the way up the scale to the VW Passat, this toolkit would be a better choice, with the one-off costs in the billions being spread over a large number of cars. There was another reason for the change in priorities: the Works Council prefers electrification of the MQB because the Golf and Passat using it are built in Germany – unlike the Up.

Future-proof battery sourcing

The feasibility of all plans for both hybrid and electric cars is primarily dependent on whether VW can access high-performance energy storage technology. To date, battery life and range have been limited by electrochemical issues and have often been unreliable. Furthermore, batteries are currently extremely expensive. At present, one type has a clear lead: the lithium-ion battery. The Wolfsburg engineers will have to secure access to this technology and to the cells, and have been trying to accomplish this is in a variety of ways: they have requested cooperation with the Chinese firm BYD (and currently have little more than a declaration of intent to show for it), with Japanese companies (Sanyo Electric Company and Toshiba Corporation), and with Bosch-Samsung, all the while keeping the door open to other potential partners, especially the Korean LG Chem company.

This strategy, which consists of sitting on the fence and not committing too early to a technology or supplier, is something of a double-edged sword. The advantage of it – not having to invest too early in technology that may lead to a dead end, as with Toyota's investment in nickel-metal hybrid cells, being a kind of cautionary tale in this respect. The danger, however, of this open approach is that, simply put, VW may miss the train and later might be dependent on battery cell producers who in reality are not neutral market players – Panasonic, for example, is clearly within Toyota's area of influence. "A focus of future development will be electrochemicals," VW's chief of R&D, Ulrich Hackenberg, explained. This much is clear: it would not be in the interests of the Group to place one supplier in a position of strength from which it could develop a unique market position, as happened with Intel and microprocessors in the personal computer market.

Even if the battery question can be settled, there will still be the thorny issue of how to balance in-house production and outsourcing. The sheer weight of car batteries (600 to 900 lbs or 300 to 400 kg) is reason enough to keep production, or at least assembly, testing, and finishing, near the car factory itself. The latter parts of the value creation chain are of high interest to VW, especially at the parts plant in Braunschweig (near Wolfsburg in Northern Germany), where the batteries for the e-Up will be assembled. If VW could add the manufacturing of electronic control systems to the cell assembly, as Toyota has done with the Prius, it would further secure core competences, and jobs. This is an argument that may weigh even heavier at VW than at some other manufacturers: the Group is always under political pressure and is seen by many politicians as having a responsibility to create jobs in Germany – and some of these politicians sit on its Supervisory Board (see Section 2.16 – The Price of Politics). When mixed with even higher-level, national strategic interests regarding competency in and access to future technologies, Volkswagen's importance to the labor market may see it being forced both to buy from German suppliers and to finish the batteries itself. Not that this is something that VW engineers would be particularly upset about, since they frequently complain about Toyota's often outmoded components (e.g. CVT). The problem is that, as yet, VW is not often able to offer a better alternative for a realistic price.

New mobility laboratory to test new business models

Neumann's successor, Rudolf Krebs, does not just have technological road maps to prepare, but must also find new concepts for mobility in the electric age. This means further research and development, linked to as many regions across the world as possible. R&D has to integrate utilities companies in order to find a feasible business model. Electric cars require a complex infrastructure, and this cannot be supported by automakers alone. Therefore, car makers can only spread any new technology they successfully bring to mass production by standing shoulder-to-shoulder with public and industrial partners. The German car industry, for example, must undertake far greater efforts to get the federal government involved – to date Berlin has subsidized electric cars far less than Paris, Beijing, or Washington.

The German central government has the aim of putting roughly 1 million electric cars on the streets by 2020. The second economic stimulus package passed before the 2009 national elections allocated around €500 million (around US$720 million) for investment in electric cars, mainly planned for research and development. Compared with France, which wants to put twice as many EVs on the roads by 2020 as its German neighbor, however, "this is a relatively small sum," complained Winterkorn. In view of the fact that the country's carmakers are behind the competition in terms of electromobility, the German government does not want to subsidize electric cars as such, which would amount to subsidizing French, Japanese, or US manufacturers. Instead, German politicians want to support R&D-related efforts in Germany.

Ambitious plans for the electric age

Winterkorn set 2018 as the date by which Volkswagen wants to dominate the e-car segment. He wants VW to be the company that mass-produces the first "electric car for everyone," and is aiming for VW to be making 3 percent of its sales from all-electric cars by 2018, when VW is aiming to be selling more than 10 million cars a year. That would require 300,000 of them to be battery-driven. Remaining cautious in tone, however, Winterkorn explained: "We need to wait to see how car buyers react and how big demand is really going to be."

Nevertheless, VW's timescale is tight. The company has established test fleets for Berlin, Florida, Shanghai, and Chengdu; by the end of 2011, 500 test versions of the electric Golf should be delivered to customers, and within another year, hybrid versions of the high-volume models Jetta, Passat, and Golf should be hitting the streets. The tempo will not slow after that, either: 2013 is going to be "the key year for electric cars," according to Winterkorn. Currently, plans are for the electric Golf to be on the market by then, accompanied by the electric versions of the SEAT Leon and of the new compact Up and then an electric-driven Jetta toward the end of the year. The extent to which the combined forces of Volkswagen and Porsche are planning in terms of electric cars became clear at the Geneva Motor Show 2010, where the Porsche 918 Spyder was shown: it is a 500-horsepower supercar with a plug-in hybrid system, including a battery cell that can be charged through standard plug sockets. At the unveiling, former Porsche boss Michael Macht went on record to say that "social responsibility" would become increasingly important even for sports cars. Macht's successor, Matthias Müller – previously chief strategist at VW headquarters in Wolfsburg and a close confidant of Winterkorn – has the task of bringing this prototype to the production stage at the main Porsche factory in Zuffenhausen. Porsche established a factory for this prototype, which is intended to hit the market in 2014. This is something that fans of sports cars *and* climate protection can definitely get enthusiastic about.

The result of all this, however, will probably not enthuse VW stockholders to quite the same extent, because VW is going to invest heavily in hybrids and electric engines and, as Group chief Winterkorn himself admitted at Geneva, "no car maker will make money on this kind of car in the foreseeable future." Mass production of modern batteries will be one way to keep costs down, but it is clear to everyone that even a giant like the Volkswagen Group cannot drive the switch from fossil to electric energy by itself. Car suppliers like Bosch, Continental, and ZF, known for their innovative approaches, will have to get involved in the process, and former Continental boss Neumann knows only too well that VW needs those strategic partners to support it, especially in terms of energy management and the electrification of the vehicle auxiliaries' drive system. Suppliers of course will be only one part of the solution, as the bulk of the research effort will be assigned to VW's in-house knowledge centers. The sites that have be made the lead players in this are Braunschweig for batteries and Kassel for electric engines.

Volkswagen's chairman of the supervisory board, Piëch, is behind a lot of the movement in green car technology, but is also counting on improvements to combustion engines: "We will produce a car with a consumption of one to two liters, i.e. below even the current LEV category," he said at a lecture at the Vienna University of Technology in 2010.

2.16 THE PRICE OF POLITICS

No decision of any importance can be made in Wolfsburg without the agreement of employee representatives. With more than 95 percent of workers unionized, including several members of the management board (such as Labor Director Horst Neumann), the IG Metall Union is a powerful combatant in the constant battle of interests at Wolfsburg. No Governor of the State of Lower Saxony has ever been able to get away with doing anything that might endanger the 90,000 jobs at Volkswagen AG – 75,000 of which are in his or her home state – or the many thousands more at suppliers dependent on the giant Group.

Participative management in Germany

Among the particular features of German corporate structures, employee participation (expressed in terminology like "Works Council" and "participative management" and otherwise known in German as "co-determination" (*Mitbestimmung*)) is a part of everyday life enshrined by law in this export-oriented economy.

The origins of Germany's unique participative corporate model date to the immediate post-war years. German industry was democratized in the years after the Second World War, with the British occupation forces obliging large companies in primary industries (e.g. mining, steel) to split their supervisory boards 50–50 between shareholders and employee representatives from 1947 onwards. The provisions of this decree were written into Federal German law and remain in force today.

Legislation passed in the 1950s and 1970s extended supervisory board parity between workers and owners beyond the core industries to all incorporated companies with more than 2000 employees. As of 2004, the threshold was further reduced to a staff of 500, albeit with a reduced quota of one-third of the seats on the supervisory board allotted to employee representatives.

Representation is structured through the works council (*Betriebsrat*), whose members are elected by employees and given four-year mandates. All companies employing more than five people on a regular basis must allow employees to form a works council. Works councils have a legal right to be comprehensively informed and consulted on all matters affecting pay and conditions; regardless of the size of the company, they have the power to block measures such as extended working hours, pay reductions, or redundancies.

In companies with 2000 or more members of staff, however, works councils may use their 50 percent of seats on supervisory boards to veto decisions taken at an executive level. While they cooperate closely with unions, works councils remain separate from them and may not order strikes. Yet their right to veto can be an exceptionally powerful tool in larger companies.

VW's employee representatives use their agreement as a strategic tool, which explains why the Brussels plant, clearly superfluous but employing 2200 workers, was kept alive. In 2007, it was taken over by Audi, which promptly invested €300 million (US$430 million) and started producing the subcompact A1 there. Another example of union power is Auto 5000 GmbH, a company which was deliberately set up by the Group outside of existing pay agreements in order to save money, but was later re-integrated again. Auto 5000's employees finally came to enjoy increased pay and benefits, after all. Instead of "slaughtering sacred cows," as former Porsche chief Wiedeking once demanded, the Volkswagen Group is rather inclined to create new ones.

The chairman of the supervisory board, Piëch, is a staunch defender of employee power in Wolfsburg: after all, he has the unions to thank for remaining at the top for so long. His opponents had been close to kicking him out after a series of sex scandals involving a number of VW personalities and allegations of corruption in 2005 and 2006. Throughout the summer of 2005, for example, the newspapers were full of stories about the then Works Council chief Klaus Volkert's Brazilian mistress (kept using a company account), the "pleasure trips" taken by other Works Council bigwigs (also paid for on company money), and the search for accounting black holes within the Group. Volkert resigned suddenly and was later convicted in court, as was the Labor Director at the time, Peter Hartz. These salacious and extremely high-profile scandals earned themselves an instant place in the annals of white-collar crime; in the official history of Volkswagen, however, they are not mentioned at all.

It was at this point that, for the first time since stepping down as VW CEO in 2002, Piëch was seen to intervene directly in management affairs. Indeed, many companions of succeeding VW CEO Pischetsrieder date the beginning of the end of the latter's career to his efforts to investigate and clear up this scandal, in which he showed scant regard for Piëch. Meanwhile, "old man Piëch" (as he is referred to within the Group) managed to keep his place as chairman of the supervisory board in the face of resistance from powerful quarters: the State of Lower Saxony, for example, tried to remove him as chief supervisor, but ran up against Jürgen Peters from IG-Metall, with whom Piëch has had a stable alliance for years.

Piëch returns the favor by publicly advocating the interests of VW employees, and goes on record saying that "it is entirely possible to run a company successfully while involving employee representatives in decisions" because, at the end of the day, there is enough effort and cooperation on all sides. There is no company of comparable size in Europe in which participative management plays such an important role. Without employee agreement, no new factory can be built, no old site closed. This is because decisions require a two-thirds majority on the Board of Supervisors, where employee representatives hold half of the voting rights – by law. This unique regulation has been enshrined in VW's articles of association, or what is known in the German judiciary as "the Volkswagen law," with which the company has to comply, even after the proposed merger or combination with Porsche.

There have been many disputes about the Volkswagen law, not least because it assigns Lower Saxony two seats on the Supervisory Board of VW. The European Court of Justice tried to remove this provision from the Volkswagen law, but both the Porsche and Piëch families represented in the Group have managed to cement the power of the federal state in VW's articles of association. The blocking minority has been fixed at 20 percent – or almost exactly the number of votes which the State of Lower Saxony controls. Then again, this blocking minority is useless when the Board is voting on closing old sites or opening new ones, since here the two-thirds majority is the deciding factor – yet another part of VW politics anchored in the VW law. "This provision has never been questioned, and it is the cornerstone of our participative management system. The Volkswagen Law can and will continue with this clause," stressed Works Council chief Osterloh, whose power, of course, rests on two-thirds majority rule.

What Osterloh is also keen to stress is how responsibly employee representatives use this instrument. Furthermore, he emphasizes that the Works Council has never blocked any projects outright, but simply examined their effect on the structure of the Group. As a recent example he points to the new production facility in Chattanooga: the Works Council in Wolfsburg was not concerned just with its effects on German sites such as Emden and Zwickau, but also on the factory in Puebla, Mexico. "New products and production facilities should not negatively affect existing sites," Osterloh states.

The way the Group conducts itself is very much a result of the provisions of its articles of association as they were set in 1960 following the privatization of the until-then completely state-run company. The Volkswagen law that put the company into private hands was full of special vetoes for the Works Council, the unions, and the Federal State of Lower Saxony. The influence of this trio makes the Volkswagen Group *de facto* a politicized company, and this has far-reaching consequences. If it comes down to it, employees can block any cuts they consider to be too harsh. International divisions of Volkswagen are brought into this culture by their own International Works Council, of which Volkswagen must also take account. Above all, though, German production sites such as those at Braunschweig, Kassel, and Salzgitter, due to the sheer number of jobs involved, are impossible to shut or sell.

Volkswagen is defined by constant dichotomy. In this formerly state-owned Group in which the State of Lower Saxony still holds a stake, conflicting groups must be integrated and allowed to coexist. There is no comparable company in which the pressure to resolve conflicting interests is so great. This continual effort to gain consensus can at times cripple this gigantic Group, or allow corruption to flourish. Then again, when everyone is headed the same way, this giant hits its stride and can move mountains. "There are lots of different interest groups, starting with the Works Council, the city of Wolfsburg and, of course, the Federal State of Lower Saxony as a shareholder who can make decisions more difficult," says CEO Winterkorn, who nevertheless is of the opinion that everyone is in agreement about the direction the company must take. However, the test of the change in this balance of power, which will include the Emirate of Qatar with its expected two seats on the Supervisory board, has yet to come.

As of 2006, both profit and securing jobs have equal priority as goals of the Group; during the year of crisis preceding this, there was mutiny in the air. "It was a thoroughly difficult situation," confessed the current Labor Director Neumann, looking back. In 2005, industry observers had branded Europe's number-one car maker a "basket case," since Wolfsburg's accounts showed a turnover of €95.3 billion (over US$137 billion) but only a wafer-thin profit of €1.7 billion (approx. US$2.45 billion). The then CEO Pischetsrieder responded with a sweeping program of cuts and savings. He cut 14,000 jobs and ended the four-day-week practice – without increasing wages – which has since saved the company at least €1 billion a year. The agreement with IG Metall specifies an upper limit of 34 hours per week and defines concrete production targets for each manufacturing site in order to ensure greater productivity and to secure jobs.

The employees agreed to this program, and in return, management signed a jobs guarantee for the 90,000 core staff of the Western federal states; an extension through to 2014 has already been agreed and now includes some 8000 workers in Saxony (former East Germany) as well. This is great for the employees but difficult for management, which has the task of keeping all of these employees busy and increasing productivity at the same time (the target is set an ambitious level of 10% per year). It goes without saying that the interests of management and the employees do not always run parallel, yet both Winterkorn and Osterloh are eager to stress that they speak openly with each other behind closed doors. "It is far better to talk with each other than to write letters," says the former. In fact, there is often a high degree of coordination, as became clear in the struggle with Wendelin Wiedeking and Holger Härter from Porsche in 2008–9, when Osterloh exercised very public criticism on behalf of VW so that Winterkorn and the Management Board did not have to confront their opponents from Zuffenhausen.

Whenever there are conflicts of interest between management and the Works Council, this attitude leads to compromises that bear the unmistakable stamp of Wolfsburg diplomacy. When the employee representatives wanted to extend the bumper spraying facilities, the managers went away and calculated that it would be cheaper to outsource them. At the end of the day, management invested in extending the bumper spraying facilities, but not as much as the employees had demanded, and the scope of work was increased so that gas caps are sprayed as well. The rest is done by a regional supplier which is kept alive thanks only to the Volkswagen orders it receives. Since the mid-1990s appearance of controversial purchasing manager José Ignacio López, however, purchasing adheres far more to the rules of competition, and price has become a decisive factor. If the quality is not as required, then an emergency expeditionary force is dispatched from Wolfsburg to bring things to order.

The López era at Volkswagen

One of the most controversial episodes in Volkswagen's history began at the height of its merciless drive for competitiveness in 1993, when José Ignacio López de Arriortúa came to the Group. He had already made a name for himself as a particularly hard-nosed cost-cutter at General Motors when he was poached by Piëch to become a member of the Board of Management in charge of optimizing production and supply at Wolfsburg. The Spaniard was preceded by legendary tales about his fanatical attitude toward his work. He was known to refer to his closest colleagues as "warriors," but the new methods of cost management he soon implemented in Germany earned him a new nickname among suppliers there: "The Strangler". His way of` working brought bottom-line results that no one in Wolfsburg would previously have thought possible. It was López who was substantially responsible for moving the Group to the highly efficient just-in-time practice by which components are not delivered to the line until the point in the process at which they actually have to be mounted, and he reduced manufacturing depth by a great deal. This forced Volkswagen's suppliers, not Volkswagen, to shoulder investments to increase productivity.

Yet López was not able to concentrate on cutting costs for long. His former employer GM accused him and the team he took with him to Wolfsburg of industrial espionage. Crates of documents and data disks were said to have been shipped from Detroit to Wolfsburg, detectives were employed to find proof, and GM and its German division Opel brought charges against Volkswagen in its home country.

At the same time, GM also went after Volkswagen in the USA. Using RICO legislation passed to fight the Mafia and other organized crime, GM attacked Piëch, who had publicly defended López from the accusations. It took two heads of state, German Chancellor Helmut Kohl and US President Bill Clinton, to reach a compromise under which López was forced to step down in 1996 and VW paid US$100 million in damages to GM and agreed to purchase parts worth around US$1 billion from the American company.

López may have left, but the methods he used to cut costs stayed; in fact, things got even more radical after his departure. While López and his "warriors" did not simply demand lower prices from suppliers, but also suggested methods and areas where savings could be made, their ambitious followers in the purchasing departments of almost all vehicle manufacturers decided to reduce the effort put into meaningful calculations, and simply took to making absurd demands of suppliers. Frequently, second-class purchasing managers rebranded themselves as "cost-cutting gurus" and, under the aegis of the manufacturers, squeezed the suppliers like lemons; this attitude had serious consequences for parts suppliers, some of which were driven to bankruptcy by it.

In order to be successful in the long run, Volkswagen must find an answer to one very pressing question: what is the core competence of a car manufacturer? This means asking what the Group must do itself in order to secure competitive advantages like the one it has in terms of dual-clutch transmissions, and what can better be handled – and bundled – by suppliers.

New ways of thinking will not affect just standard parts such as seats or fenders; the transition to next-generation cars, for example, poses questions about the value chain as it currently exists. To date, manufacturers such as Volkswagen have been producing engines in-house in order to guarantee their models' individual character, quality, and driving "feel," and to create value. However, once mechanical drivetrains, i.e. combustion engines and gears, have been replaced by electric motors and high-performance batteries, car manufacturers suddenly face an enormous challenge – especially manufacturers like VW Group that have invested so heavily in manufacturing engines. The Group now, after no small amount of hesitation, is turning to electric cars and is looking for ways to increase efficiency or release capacity in order to finance the not inconsiderable investment in battery-cell technology and super-light construction needed for this to be a success.

The employee representatives also want the Board to raise external capital for products outside of the Group's core automotive competences. The first such initiative started in late summer 2009 with combined heat and power (CHP) gas engines being now produced for domestic use. Moreover, VW's management is considering adapting VW engines for agricultural or construction vehicles, mowing machines, or wheel loaders. In May 2010, the car maker produced a study called "Bike.e", a bicycle with an electric motor attached that can reach speeds of up to 15 mph (25 km/h) and can be folded down to a package the size of a car tire to fit in the space for spares. The motor can be charged over the in-car cigarette lighter and, once loaded, has a nominal range of 20 km (12.43 mi).

Yet this is not enough for the employee representatives. In mid-February 2010, as IG Metall was negotiating on extending the jobs guarantee for the six western German plants, VW's financial services branch, and the plant in Saxony, it also took the opportunity to agree with management on a fund for investing in new areas of work that goes above and beyond the resources already earmarked for innovation in the annual budget. This means that, along with an innovation fund already in existence, Volkswagen now has an eight-figure sum to spend on new fields that should guarantee work for the Group's employees in the future. Management and employee representatives consult on how to disburse the resources made available.

Works Council chief Osterloh is pressing for the Group's at least 20,000 engineers to be allowed to work on products besides cars and is going public to demand concepts for "products outside of the automotive value chain" in order to find use for capacities and guarantee employment in the long term. Yet CEO Winterkorn refuses whenever the sums get too high, pointing to Daimler's failure when investing outside of its sector in the German-based electronics manufacturer Telefunken, which at the end of the day had to be dismantled and terminated, or to Volkswagen's misadventure with the German-based computer manufacturer Triumph-Adler. Instead of such ventures, the CEO is aiming to increase internationalization in order to keep the plants at home busy. With the launch of the Audi A1, for example, Winterkorn sent a message to his employees: "We can only keep our factories at home running if we produce successfully on the world market." He pointed out that the model produced in Brussels was equipped with numerous parts from Germany, and added: "and that's another way of keeping jobs at home."

In Wolfsburg, it is the technicians and engineers who have the say. While experienced sales personnel might be highly appreciated, they would not really be considered CEO material. Head of Finance Pötsch has played an important role in the Porsche takeover, and he makes sure the strategists have enough money to play with, but there will be no businessman at the very top of the Volkswagen pyramid anytime soon. "A carmaker must be run by somebody who knows the product," Winterkorn said, echoing the words of his chief supervisor Piëch. Enjoying car driving was simply not enough, according to Winterkorn; the CEO must understand how a car is put together. Furthermore, VW saves time and mistakes because no one can pull the wool over his eyes: "When my engineers tell me that something won't work due to technical, temporal, or financial constraints, I am perfectly capable of arguing against them, and everybody is aware of that," he stressed.

Nevertheless, there is no shortage of people who see the accumulation of roles around the CEO as a bottleneck: Winterkorn is not only CEO of the entire Group, but also leads the core VW brand, Porsche Holding SE, and Scania's Supervisory Board. Nothing substantial can be undertaken in Wolfsburg without Winterkorn, but he simply cannot be everywhere at all times.

While this slightly authoritarian side of the Wolfsburg management culture does provide a clear line of demarcation, it also tends to damage the self-confidence of second- and third-tier managers when major decisions are taken at top level – which is particularly the case when a brand shows signs of weakness. Winterkorn then puts its managers on a leash, but when brand managers are no longer allowed to make decisions, they tend to become frustrated and less dedicated. CEO Winterkorn, whose angry outbursts when faced with problems and errors are feared internally, does not take kindly to signs of slacking: "It is crucial to carry our product philosophies forward," he said, "because being ahead of the field is not enough." In his view, complacency is the danger, and success must be continually justified. The management job Winterkorn took on in November 2009 after the Porsche takeover and the integration of the sports-car maker into the Group has made his to-do list considerably longer, and the most likely item to be squeezed in his already very tight schedule is the core brand, Volkswagen. This is an ever-present danger for the Group's progress.

It is a danger, however, of which the Group seems aware – and which is currently being mitigated by the astonishing recovery in worldwide car sales in 2010, as well as the unexpected weakness of the Group's rival Toyota. Yet mitigated does not mean banished, and whilst cost-cutting in the core brand has gone on apace in recent years, there is no shortage of opportunities to save costs that have so far gone unused. Volkswagen may no longer be quite the "unpolished diamond" of which former manager Bernhard spoke, but it is certainly still no shiny showpiece. This potential has been visible in the past, such as during the Bernhard years when productivity was noticeably improved, and is at its clearest today when Winterkorn shows his visionary side in relation to products. Were these two approaches to be consistently and stably linked, they could well catapult Volkswagen onto another planet in terms of profitability. Yet it is highly questionable whether the political currents in the Group would ever flow so as to allow such a match-up, and so the new shine on the Volkswagen diamond is in no small part due to the bright light of the current economic situation. If this light goes off again, as it did in 2009, much of Volkswagen's new gloss will lose its luster. Many economists are predicting a double-dip recession in which 2010 and 2011 will have been two deceptively sunny years sandwiched in between two very severe economic winters. In this scenario, the Volkswagen core brand especially would need to cut costs even further, but it is hard to see how this could be done without breaking the political taboos that are so crucial in the Group; the currently calm climate at Wolfsburg could give way to a storm, and company strategists and politicians will once again be at loggerheads. For Volkswagen, strategy and politics can interact both to the benefit and at the expense of the profit margin, as is the case of the insolvency of the German-based car-parts supplier Karmann. Volkswagen bought the bulk of the production facilities, which, to the extent that it will secure jobs, was very much more in the interest of one particular shareholder – the Federal State of Lower Saxony – than in that of the Group itself. Nevertheless, there is a hidden opportunity here: Volkswagen has spent just under €40 million (US$58 million) on the site, which includes a modern paint shop that would cost more than €100 million (or US$144 million) new. The Group is now starting to produce future niche models like the Golf convertible at this site near Osnabrück in north-western Germany. Additionally, the former Karmann site is earmarked to manufacture Porsche's Boxster and Cayman models, if sales pick up. Hitherto, contract manufacturer Valmet of Finland has produced those models, although the bulk of the manufacturing is done at Porsche's headquarters in Stuttgart, Southern Germany.

Yet politics does not always come this cheap. In terms of raising productivity or solving other fundamental problems, Volkswagen's hands are politically tied. The fact that the Works Council and the Federal State agreed to the €16 billion (over US$23 billion) takeover of Porsche AG and its sales holding means that chief supervisor Piëch must take extra care to get along with them in the coming years. Both parties will expect repayment for their loyalty at some point, as will the Emirate of Qatar, which dug Porsche Holding SE out of a billions-deep hole of debt.

This constellation of mutual interdependence makes it even more difficult to significantly cut costs at home plants, and it remains to be seen how the refashioned and significantly larger Group will stand up to any further turbulence in the automotive sector. In such circumstances, every percent of perfection that VW trades in for productivity, quality, and profit would help in difficult scenarios. Indeed, it is the sheer lack of discipline in keeping internal costs down that is the most noticeable inconsistency of the Volkswagen Group and its chief supervisor Piëch, who himself admits that he "dislikes coming in second" and, interviewed in 2008, that he was annoyed that Toyota was ahead of Volkswagen for "no good reason," as far as he could see. "It can't go on like that, of course, but I am motivated to change it by a feeling of wanting to be better, not by feelings of anger – anger is not a good motivator in the long term," Piëch said. Now, 2011 may well turn out to be the year in which Volkswagen actually does pull ahead of Toyota, and whilst Piëch will have every cause to be satisfied with the work he and everyone in the company have undoubtedly put into achieving this, there will be those who will say that it is now Volkswagen who is ahead for "no good reason". Many observers see Volkswagen as having disproportionately profited from factors beyond its control, especially the strong economic climate in its core markets Germany and China as of 2010 and the tragic catastrophe which struck at Toyota's home country in March 2011. What these observers tend to point out is that those who live by the sword tend also to die by it, and that those who profit from external factors can also be toppled by them. An example is Volkswagen's strong position in China: it is paying off at the moment, but China is not a free market economy and Volkswagen is in the final analysis at the mercy of the Communist party; in a changed political climate, it may suddenly not be so advantageous to be selling a quarter of the Group's output there. Due to the importance of China to VW the Group's approach to EV – a key aspect of its future – is now in no small part dependent on how the Chinese government choses to subsidise electric mobility, and to what extent it pushes Chinese-only solutions. Just as Toyota is highly dependent on the US market, Volkswagen's fate may well be decided in China. Such is the scale of the risks which the Group's current success is currently obscuring.

In any case, if Volkswagen falls out of favor in China, or puts a foot wrong in future technology, or if the economic growth on which it is currently thriving melts away, Toyota will be waiting to step back into its old shoes as number one. Although the firm has experienced shock after shock in recent years, it is not going anywhere soon: Toyota has huge financial reserves and a company philosophy based largely on progressively cutting costs and increasing productivity. Volkswagen's executives, on the other hand, have to struggle to impose some measure of discipline on an organization which does not have a culture of it; and imbued as they are with this culture, they – like Piëch – can often be found wanting it themselves.

2.17 PIËCH, THEN WINTERKORN – AND WHO NEXT?

Piëch is a classic patriarch, and during his reign whole generations of management men have come up against him – and gone away licking their wounds. Men such as his deputy when he was Chairman of the Board, Daniel Goeudevert, the ex-Audi bosses Herbert Demel, Franz-Josef Kortüm, and Franz-Josef Paefgen (later boss at Bentley) and most recently – and most spectacularly – the Porsche boss Wiedeking. Piëch knows no mercy when it comes to getting rid of people but is generous in the severance packages he hands out. "It is often cheaper to pay a full salary for four or five years than to leave the wrong person in a top position, but it has to be legally and morally correct, of course," said the man himself.

Piëch corrects mistakes, too. After Wiedeking was forced to go in July 2009, Porsche was lacking a strong leadership personality. Wiedeking's internally determined interim successor Michael Macht calmed the atmosphere after the failed takeover of VW, but both the managers in Wolfsburg and those at Porsche's headquarters in Stuttgart wanted clear leadership. Much of Porsche's middle management thought Macht too hesitant, yet his grasp of production matters stood beyond question. For this reason, the VW grandees "poached" him from Stuttgart and, showering him with praise, used him to replace the luckless Head of Production at the Group level, Heizmann. This is in many ways a demotion disguised as a promotion. The successor to the Porsche throne, Matthias Müller, has been with Winterkorn since his Audi days and is an acknowledged member of the inner circle.

People who have known Piëch for a while attribute a lot of his ability to get things done to the fact that, compared to many other billionaires and several members of his own family, he chose a career in the automotive industry and has often been forced to remain strong in the face of resistance – his status as heir at Porsche, Audi, and Volkswagen was in many ways a burden rather than a boon.

Winterkorn is also no stranger to adversity, although very much in the other sense, coming from an austere and not particularly wealthy background in the devoutly Protestant region of Swabia, Germany, and then working his way to the top. The two car fanatics first met at the beginning of the 1980s. At the time, Piëch was Management Board Member in charge of Technology at Audi and took on Winterkorn as an assistant for quality assurance. They both took a language course called "Spanish while you sleep" and had plenty of opportunity to talk shop, and it was during these conversations that they realized how similar is the view they share of the automotive industry. When Piëch took the throne at Audi in 1988, he made Winterkorn his Quality Assurance Manager. Five years later, the Porsche heir had reached his goal and was leading the Group in Wolfsburg, and Winterkorn was made responsible for vehicle development. In view of the closeness between the two men, it is no wonder that, when he retired from the job in 2002, Piëch toyed with the idea of promoting Winterkorn directly to Chairman. Instead, he sent the younger man down to Ingolstadt to manage Audi. Winterkorn was surprised by this decision and would much rather have remained Head of R&D under the new VW CEO Bernd Pischetsrieder.

In hindsight, it was entirely correct to send Winterkorn to gain experience as head of Audi, says the man himself: "It would have been far more difficult to go directly from Head of R&D to CEO. At Audi, I was able to gather knowledge in important areas such as finance while learning how to stand in the spotlight in comparative calm." People who know him agree with Winterkorn: before his time at Audi, his temper – which frequently gets the better of him and leaves him speaking so fast that he becomes difficult to follow – lacked the polish needed of a head of the Group. On a personal level, Pischetsrieder and Winterkorn got along fine, but their opinions on how to divide up work were fundamentally different. There was a constant tug of war between the consistently self-confident subsidiary in the South and the Group center at Wolfsburg in the North, and Winterkorn was anxious to stake his claim to independence and its trappings. While Pischetsrieder and his Chief Design Officer Murat Günak were planning to present the full range of Group innovations at the Geneva Motor Show of 2002, including the Audi A6, Winterkorn as Audi boss and his Italian top designer de'Silva threatened outright to not go to the Motor Show at all; they got their way and presented the vehicle themselves. Winterkorn is very clear on this point: "The presentation of a new vehicle must be done by those directly responsible for it, those who are passionate about the cars they are producing."

Audi certainly profited from Piëch's decision. By that time, they had lost the taste for attacking BMW and Mercedes. Winterkorn brought just the jolt the automaker needed, and while his relationship to HQ in Wolfsburg remained tense, he used his position as the new boy in Ingolstadt to make extensive use of the expertise of patriarch Piëch. Whatever the car, whether it was a twelve-cylinder engine or the 600-horsepower "Porsche-killing" supercar R8, as Audi boss he needed a critical "sparring partner, and Piëch was just the right man for the job," Winterkorn recalls. The constructive dialogue between the two experts paid off. The aggressive new Audi design, first shown to the public in the form of the Nuvolari showcar at the Geneva Motor Show 2003, did well and has been a foundation stone for Audi's success in the years since. This period also served to bring Piëch and Winterkorn even closer to one another. Porsche heir Piëch relies on a very small circle of trusted insiders and none more than Winterkorn. In this, Piëch is very much following the leadership style of his famous grandfather Ferdinand Porsche (whose first name he shares). The car-building genius gathered a team of experts around him who dealt with the details the way this perfectionist wanted his vehicles finished.

The two men at the top of Volkswagen are connected by their love of cars: "We love cars: both of us go up to automobiles and stroke them – really stroke them. Only a few people would do that," says CEO Winterkorn. The extent to which both VW grandees are still bedazzled by cars is shown by an anecdote that made the rounds a while back. In the summer of 2008, as the pre-series version of the current Golf VI was just being finished in Wolfsburg, an important meeting about the future of Porsche was taking place in Salzburg, Austria, where the Piëchs come from. Many members of the hopelessly estranged Porsche and Piëch clan were there. Winterkorn, who was also taking part, had one of the first Golfs brought down to the sun-drenched mountain terrace in Salzburg – as always painted in black, he explains himself: "We have a tradition: Piëch and Winterkorn are allowed to drive the first two models of any new car." During the often heated debate about the fate of the financially desperate Porsche company, the VW chief supervisor and Porsche heir Piëch could not keep his eyes from wandering longingly to the car outside the window of the family property, cooing about what a beautiful model it was.

Piëch's supremacy is one of a kind in German industry, and makes it difficult to answer questions about how the future of the Group will look once he takes his leave. Winterkorn is clearly readying himself to take over as chairman of the supervisory board, but at the moment he is needed as CEO, and will be until the end of 2016, at which time he will be 69 years of age. Piëch, meanwhile, has made it known that he wants to continue until the 2012 General Shareholders' Meeting at the very earliest. Whenever he goes, however, there will be a gap somewhere. "We will have to prepare people to lead the Group," says Works Council chief Osterloh, "and in order to do this, we need to identify the right successors. We currently have an excellent management team, but all of them are around 60 years old." Yet in no company is a handover of power as delicate as in the strictly hierarchical Wolfsburg system. "There is no crown prince as of yet," said Winterkorn.

The break-neck speed of the Group's expansion requires new management structures, which keep the growing portfolio of brands and models on track. Winterkorn is aware of how difficult it will be to find people who can take over. "In the next generation, we will be looking for people who have the aggressiveness needed in terms of products, technology, and quality," the CEO stressed and made known his criteria. As far as he is concerned, individual personalities can make all the difference to the Group:

> The whole company is trimmed to suit the team and the man at the top: The management team is there to convince the others that they are right. Success and failure at Wolfsburg has a lot to do with who is leading the company. This is definitely a challenge in view of the transition to the next generation.

Rival Toyota, meanwhile, has already passed on power to the heir of the founder – who had to overcome the most painful start-up difficulties in the first two years of his reign.

Interview with Karl-Thomas Neumann,
Volkswagen's Chief of Chinese Operations (November 28, 2010)

INTERVIEWER: In 2010, Volkswagen sold almost 2 million cars in China, but won't the explosion in demand will make it increasingly difficult for you to keep your market share of 18 percent?

NEUMANN: It certainly will not be easy, but we are optimistic about our chances of achieving this in 2011. What will be equally challenging, however, will be keeping quality at the right level in view of the speed of growth. One of my main areas of work is avoiding operational risk where at all possible. We are building new factories and developing new models in the Chinese market and, in order to make sure this comes off without serious errors, we need to keep increasing the size of our staff.

INTERVIEWER: What does the Chinese government's acknowledged aim of developing domestic automotive manufacturers mean for foreign companies in China such as VW?

NEUMANN: It means that we need to show precisely how much we have contributed to the development of industry and society in China over the course of the decades. Up to this point, Volkswagen has invested €1.9 billion in R&D in China, and this sum is now set to grow considerably – we have to communicate this sign of our commitment and then continue to be a good corporate citizen. My central aim is for us to still be able to talk about the future of Volkswagen in China ten years from now.

INTERVIEWER: What is going to happen in the mid-term?

NEUMANN: The overall trend remains an upward one. Currently, there are about five cars for every 100 people in China, whereas there are already 50 cars on the road for every 100 Germans. Meanwhile, the Chinese middle classes are growing and looking for status symbols such as automobiles on which to spend their disposable income.

INTERVIEWER: What are you going to do with the €10.6 billion slated for investment through Volkswagen's Chinese joint ventures through 2015?

NEUMANN: We need to invest in capacity. At present, I see no prospect of the situation in our factories easing, so we are constructing four completely new automobile plants. One of these new facilities will be located in the west, a region to which Volkswagen is a newcomer. We are going on attack in the south, too, where another of our new factories will be going up. At the present time, the south is a stronghold of our Japanese competitors. In general, our Chinese strategy is becoming more and more regional, and this is a big step.

INTERVIEWER: Meanwhile, however, Volkswagen is also aiming for clear increases in the sales of imported models in China. Are you confident of success here?

NEUMANN: Through October 2010, we increased our total imports by 130 percent to more than 60,000. Sales of premium vehicles are shooting up, so we are doing very well on our top-of-the-range Phaeton model. The current waiting time for one is nine months, which shows just how much our Chinese customers value our workmanship and handmade products.

INTERVIEWER: Are customer attitudes changing noticeably?

NEUMANN: Yes, very much so! Increased experience leads to increased demands, and many Chinese customers are now already on their second car.

Nevertheless, Skoda has high potential as an entry-level brand in China, and this was shown by our successful market introduction: in the first nine months of 2009, we had 60 percent growth as against the same period of the year before. Moreover, the new head at Skoda, Winfried Vahland, is my predecessor here in China and, as such, has some great ideas for new models.

INTERVIEWER: Competition is heating up fast, though.

NEUMANN: Yes, General Motors is a particularly strong competitor in China, and Hyundai, too, is one to watch. Then there are the Chinese companies, of course, and I think that the state-run SAIC will do very well in the years to come.

INTERVIEWER: It seems probable that they will concentrate on electric vehicles. What will Volkswagen be doing to counteract this?

NEUMANN: Electro-mobility will play a huge role in China in the coming years, and both joint ventures in which Volkswagen is engaged will begin selling their first mass-market electric vehicles in 2013 or early 2014. As manufacturers, we have the challenge of keeping tight control of costs, so we hope that volume will grow quickly. Above all, we will not develop any electric motors that do not fit into the modular toolkit system, as this is the only way that we can achieve the kind of costs scaling we require for electro-mobility to be profitable.

INTERVIEWER: How does China specifically fit into this plan?

NEUMANN: Our engineers in China will be charged with developing good-value modules for the transverse toolkit, including components specifically produced for electric cars; one of the most important things will be a battery that fits the toolkit system. We will be buying our cells from Chinese suppliers and then developing our own

Volkswagen battery architecture, electronics, and cooling system. The cells will have to be standardized in order to keep costs down.

INTERVIEWER: Why can't Volkswagen make a new model under its own steam and market it as the pioneering electric vehicle – like Apple did with the iPhone?

NEUMANN: A completely new design, with a sleek, new body made of super-light materials is a real temptation for both our technicians and for our marketing department, that's for sure. But this kind of extravagant approach is simply beyond our financial means. Besides, the Chinese translation of Volkswagen – "car for everybody" – is very close to the German original, which means "the people's car," and so our aim as a company must be to produce electric cars that people can afford.

INTERVIEWER: What is the role of financial incentives in China?

NEUMANN: The extent to which electric cars succeed on the market is very much a question of political frameworks. The Chinese government has declared that it aims to have 5 million electric cars on the road by 2020 – and provided that it offers enough incentives, this will be easy to achieve.

3

TOYOTA: AN AUTO GIANT OVERCOMING A GIGANTIC CRISIS

As we move up to the date of the 'Toyota restart', February 24, the date of the U.S. Congress hearings one year ago, I want to ask each of you to take practical actions in the initiatives [we began last year for restoring customer trust]; specifically, to promote 'The Toyota Way with a face'.

—2011 New Year's Message for Employees from President Toyoda (Available at http://www2.toyota.co.jp/en/announcement/110114.html)

On January 14, 2011, when Toyota CEO Akio Toyoda made this request in his new year's message to his employees worldwide, he could in no way have guessed at what 2011 held in store for Toyota. Yet despite the nuclear inferno at Fukushima and its effects on Japan, or the unprecedented sales crisis two years earlier, it was the recall disaster that transformed the struggling giant in the most fundamental and most lasting way. Before we analyze Toyota's current woes, however, and its chances for staging a comeback to the top spot of the industry, we need to take a closer look at the transformation which saw a tight-lipped, centralistic Japanese company become a more open, decentralized and responsive organization; an organization whose CEO now regularly appears in the USA and other foreign markets to present Toyota's new face and corporate mission.

It was on February 24, 2010, that Akio Toyoda, grandson of the company's founder Kiichiro Toyoda, appeared before Congress in Washington, D.C. to testify on the 34 deaths that – up to that point – US authorities had been investigating with a view to a causal link with defects in Toyota and Lexus cars. It was the day he would later call the turning point in more than 70 years of Toyota history.

As Toyoda steps out into the storm of camera flashes in front of the assembled members of the House Oversight Committee of the US Congress, it is not just the legacy of the Japanese Toyoda family (an industrial dynasty the Japanese commemorate with almost religious adoration) that he had to defend. In the weeks before this memorable testimony, the disastrous product recall had already become part of a wider international dispute about trade deficits and imports, and anti-Japanese sentiment had been boiling over in the American media. In Japan, however, there was no shortage of people who saw the recall scandal as overblown and considered that there was no way of proving that the vehicles involved had really had defects. Yet the United States of America, still holding a majority stake in General Motors at the time, as well as a double-digit stake in Chrysler, was no longer a neutral party in the argument about sticking gas pedals and floor mats that were supposed to have caused the deaths. Toyota, the company that has been squeezing the US market share of General Motors, Ford, and Chrysler for years now, still produces almost every second car it sells in the USA at home in Japan – and it is this, among other things, that Akio Toyoda has to defend against concerted attacks from US congressmen.

Yet Toyoda had not come to Washington just as the man at the helm of the biggest carmaker in the world. He is also "one of Japan's favorite sons," a high-ranking representative of "Japan Inc.", the business model that turned the rocky island nation into one of the world's most successful export economies. "Made in Japan" is, after all, a brand of sorts that used to be synonymous with high quality and reliability. Now all of this seemed suddenly to have been put into question, as years of trust apparently had just been thrown out of the window. In particular, it was Toyota's image as a leader on quality that had suffered after US authorities forced the carmaker to start the largest-ever continuing recall in automotive history in the wake of a high-profile accident involving a Lexus that led to four deaths. In the wake of the Lexus accident, Toyota had to call millions of vehicles back into service centers at dealerships to check gas pedals and brakes – and the embarrassment did not stop there. Throughout 2010, further recalls were made necessary to address ever more defects, with more than 11 million cars checked to date (for a detailed documentation see the chronology in Section 3.10). Just a few months before that memorable February day, such a massive recall would have been absolutely unthinkable for the industry's biggest manufacturer, one that has always explained its success both to the world and to itself in terms of superb quality. Toyoda's job on this day was to publicly retell this story of quality, as well as to make sure that nothing speaks against it; this will remain part of his remit for the rest of his term at the top of the Japanese company.

As he was led like a criminal into the room filled with journalists staring at him and congressmen eying him critically, the media-shy chief who, until then, had only had a few opportunities to gain experience in handling major crises, was facing the most difficult moments of his leadership. In order to win over disquieted Toyota employees and the Japanese public, he would have to successfully defend Toyota against the serious charges – *viz.* that the manufacturer had known about defects in its vehicles and covered them up, had put human life in danger for the sake of profit. At the same time, he knew that it was crucial to avoid anything that could turn the smoldering trade dispute between Japan and the USA into a full-blown war.

Beginning with a prepared speech in English, Toyoda showed an unusually high degree of emotion for a Japanese CEO. He was almost shouting as he stressed how seriously Toyota takes quality and that the manufacturer would never try to cover up quality issues. He preached Toyota's sacred precepts of *genchi genbutsu* (or "go see for yourself")[1] and "developing quality people."[2] At the end of this speech, for which he had been trained by experts, he got exceptionally personal: "My name is on every car!" he said, repeating it almost defiantly as if the American audience before him ought very well to understand that Toyota, in contrast to carmakers like GM or Chrysler, embodies the entirety of a nation's proud industrial heritage and has not become the largest carmaker in the world by chance. What followed was a staged question-and-answer duel of the courtroom sort in which the two sides seemed to speak entirely unrelated languages – and not just because Toyoda was now relying on an interpreter to make himself understood. He was flanked at all times by Yoshimi Inaba, the sharp Head of Toyota USA, who made every attempt to support his boss whenever the questioning got intense.

Opinions about Toyoda's performance differed widely after the event. In the USA, he was accused of having dodged Congress's questions. Meanwhile in Japan, the hard and, by East Asian standards, often rather rude questioning caused some outrage. Toyoda himself later went on air in America questioning whether he was given an entirely fair hearing; even after three hours of relentless interrogation, this is about as impolite as a Japanese person would be willing to get. Yet Akio Toyoda did say something else on February 24, 2010 that will turn it into a crucial date in the annals of Toyota: he publicly stated that he thinks the company may have grown too fast in recent years. This had to suffice. He was not willing to offer more of an explanation to the millions assembled in front of their televisions on both sides of the Pacific. Nevertheless, Toyoda's statement lay one truth bare: regardless of how far the causes and effects of the recall are actually due to any quality issues, beyond all shadow of a doubt the huge carmaker is dealing with the consequences of a very long period of rapid growth – the undesirable side effect of which culminated in the congressional hearing in February, 2010.

By early 2011, the economic crisis of 2008–09 and a larger-than-ever recall series had weakened the Japanese carmaker, and at just this moment, the Great East Japan Earthquake struck. The impact of the subsequent tsunami dealt the manufacturer yet another heavy blow, disrupting supply chains and sowing fear and confusion. In the last three years, the company has been presented with three crisis situations of a high order – and this unholy trio at a time in which Toyota has been called out by Volkswagen to a one-on-one race for leadership of the industry – a race which, until the end of the economic boom in 2008, Toyota thought it had already won.

3.1 THE "TOYOTA SHOCK"

The company that was formerly the undisputed leader in the automotive industry has seen its strength sapped again and again in recent years. First came the global economic downturn, which hit Toyota far harder than its challenger from Germany. While Volkswagen came away with a few shots over the bow, Toyota took a full broadside, accumulating JP¥436.9 billion (US$4.69 billion, €3.3 billion) in losses in the 2009 financial year, and sinking deep into the red (NB: in accounting terms, Toyota's financial year follows Japanese practice, running from April 1 to March 31). To put this into context, Toyota had not had any losses to report since 1950. In the financial year 2009, however, sales shrank from JP¥26.28 trillion to JP¥20.52 trillion (US$282 billion down to US$229 billion/€196 billion down to €159 billion), making for a loss in turnover of almost 22 percent.

In order to put yen figures into an international context, all amounts in the Japanese currency in this and the following chapters are converted to dollars using a rate of JP¥93 to $1, the average figure that Toyota's accounting department has predicted for the 2010–11 financial year (for ease of reading, all figures are rounded up or down as appropriate). Nevertheless, applying average exchange rates for past years produces some distortions, due to the yen's drastic gain in value against the dollar since 2008. Some caution is thus required in interpreting yen figures in dollars or Euros, but whichever exchange rate is applied, Toyota's losses remain huge.

The Japanese, whose strongly export-oriented economy as a whole suffered badly from the crisis, found a suitable name for it: the "Toyota shock". Yet this was by no means the end of the horror story. The large-scale and long-lasting recall process, which started right in the middle of the sales crisis and lasted for at least a year, threw Toyota into a credibility crisis of a previously unimaginable scope. Even worse, the high-profile problems hit Toyota in its strongest sales market, the USA, where old prejudices about "Jap crap" that had been packed away sometime in the early 1980s were suddenly dusted off and back at the height of fashion, the revival being pushed by the media and supported by dissatisfied customers. This second shock hit the organization even harder than the global economic downturn, because "the lifeline of our company is the quality of our products," as the CEO unmistakably stated after the first meeting of the hastily established "Special Committee for Global Quality." In this committee, Toyota managers from all over the world now convene regularly in order to detect flaws in products earlier and make quality worldwide into an all-important part of the company's ethos.

However, this lifeline has been badly damaged, not just by the financial penalty to the tune of US$48.8 million that Toyota accepted in the USA, nor the cash that went into repairing millions of vehicles or refunding purchase money. Even though Toyota's once enormous capital reserves, which remained at ¥1.92 trillion ($22.35 billion, €15.6 billion) in cash and cash equivalents as of September 30, 2010, may have been depleted a little during the crisis, the worst thing about the quality issues was that in 2010, the carmaker that had once been legendary for its unwaveringly high standards became synonymous with potentially faulty workmanship.

For the first time in its history, Toyota had to look on as its credit rating was either downgraded or put on watch-lists by several agencies whose analysts feared that these problems could jeopardize the value of the brand, the carmaker's pricing power, and its market share. They gave the company which had always enjoyed high-A ratings (and was seen by some as more creditworthy than the Japanese government) a second or third best: in early March 2011, Standard & Poors downgraded the carmaker from AA to AA-. Then, mere days later, the Great East Japan Earthquake struck, and many agencies once again put Toyota on review for a possible downgrade due to the financial fallout from suspended car production. Moody's, for example, put the automaker down from Aa3 to Aa2 in late June, citing not just the effects of the Fukushima catastrophe, but also of the strong Yen and rising raw materials costs.

The company is in no immediate financial danger, however. More than anything else, though, the crises of recent years have left deep psychological scars within Toyota. The organization has always reacted to traumatic events with long-lasting changes in behavior, integrating lessons learned from crises into the basic fabric of the company philosophy. This became obvious as early as the 1950s, when Toyota experienced its first and last labor conflict in Japan. Issues that are part of the scenery for European and American carmakers – unions organizing strikes in the face of planned redundancies – led to a collective trauma of sorts with Toyota.

Labor conflict at Toyota

At the beginning of the 1950s, Toyota was heading for bankruptcy, and the company planned to make more than 1500 workers redundant. The dispute with the unions that followed lasted for three months, and while the redundancies were accepted, the entire management was forced to resign *en bloc*. It was not just Toyota that was hit by the strikes: Japanese rivals Nissan and Isuzu, for example, had also gotten into difficulties due to the postwar recession in Japan and were forced to cut their workforce. Toyota had lost over 2000 employees once all was said and done. External unions were forbidden from this point onwards, and the internal union was retooled into a subdepartment of Personnel; its new role was more to organize staff training than to try and secure wage rises.

The strike lasted for 90 days and left a deep psychological trauma that haunts Toyota to this day. To avoid any kind of repetition of this catastrophe, workers that stayed with Toyota at the time were given a lifelong employment guarantee, and even today, one of the core Toyota precepts is to avoid making regular employees redundant under any circumstances.

Toyota would not be a standard bearer among Japanese companies if it did not fundamentally examine the reasons for its loss of face and then integrate the insights gained from this self-examination into the DNA of the organization in a way that will affect, not just the next few years, but all future generations of Toyota employees.

Yet before we dive deeper into Toyota's anatomy and tradition, we should ask why the expansive growth policy on which Toyota has been running has made it so vulnerable to the trilogy of crises that have recently struck – especially to the sales collapse of 2008–2009, which its European rival Volkswagen managed not only to survive, but to storm out of stronger than ever.

The global financial and economic crisis put a stop to years of unparalleled growth for Toyota. In the ten years leading up to the crisis, the company had been on a persistent course of expansion and had succeeded in almost doubling its vehicle sales worldwide, moving from 4.7 million in 1999 to 8.9 million in the 2007–8 financial year. Toyota increased its sales by roughly 500,000 vehicles per year, a rate of growth which no other automotive company has ever achieved before, and it seemed as though there was nothing that could stop them. At the height of its power, Toyota deposed its biggest rival at the time, General Motors, and took the automotive crown when it sold more cars than any other manufacturer worldwide at the beginning of 2008 (measured by yearly sales). The newly crowned king of the automobile industry then achieved a record global market share of about 13.5 percent.

Nevertheless – with the exception of its hybrid engine – Toyota was neither particularly respected as a great innovator, nor were its vehicles in any way uniquely designed or revolutionary in terms of driving quality. Toyota simply produced reliable cars that customers liked, and sold them at prices that represented good value.

Yet it was nothing less than the entire Toyota philosophy in its much-famed rigor that led to such financial strength: production methods ruthlessly trimmed for maximum efficiency, models designed for several regions at once and a sales strategy to match, lasting partnerships with suppliers, and strict cost control. Armed with this strategy, Toyota sailed out of Japan and taught car manufacturers the world over that fear is spelled with a T. By 2007, Toyota had a market capitalization of almost US$200 billion (€139 billion), more than any other automobile company on the planet. The Japanese company had the highest productivity and highest profits in the sector: Toyota was, in short, the most profitable and dynamic mass-producer of cars, despite saturated markets and an ever hotter climate of competition between global manufacturers. Its motto back then was clearly a statement of intent: "The car in front of you is a Toyota."

Figures confirmed this: in the fiscal year 2007–8, Toyota booked an operating income of JP¥2,270,375 million, or US$22,661 million. Net income for the same period was JP¥1,717,879 million or US$17,146 million. The financial and economic crisis, however, which sent sales tumbling overnight all over the world, turned Toyota's scaling model into a losing game overnight.

This unannounced collapse in demand tore the veil from a series of issues that had simply been covered up by the years of spectacular growth. What the crisis revealed is that the manufacturer that was so used to success suffered from an overdependence on the Japanese and American markets, and it was in these markets that demand had fallen most steeply, yet here the majority of Toyota's production capacity is located. This means that, since the financial crisis, Toyota has had to fight overcapacity globally; it is capable of producing 9 million vehicles a year (without taking full account of joint ventures in China), which is roughly 1 million more than Toyota can sell in fiscal 2011 according to its own sales forecast (which, however, does not take full account of Chinese sales; including the People's Republic, Toyota sold 8.4 million vehicles in 2010).

Once the crisis had broken out, Toyota's global production and sales strategy made it extremely vulnerable. While its central production base is still Japan, with 2.9 million of the 5 million Toyotas produced in Japan in fiscal 2008 going abroad as exports, more than half of these – 1.7 million vehicles – went to the North American markets. It was in North America, however, that Toyota suffered its most bitter losses, with its sales in autumn 2008 collapsing at roughly the same rate as overall demand, namely by 30 to 40 percent. In the years leading up to the crisis, demand in the USA had been artificially increased by cheap leasing deals, which came back to haunt nearly all manufacturers in the market in the form of leasing returns and falling prices for used cars. Toyota was not immune, with its financial services branch posting an operating loss of more than half a billion dollars at the height of the crisis, including losses from exchange rate fluctuations.

Another mistake masked by success was the large number of pickups and SUVs in Toyota's product range primarily for the North American markets, and sold only there in number. It is at moments like this that the company's ambivalent product philosophy becomes obvious: the market leader understood exactly how to sell itself as a manufacturer of environmental-friendly vehicles such as the Prius, which was the first mass-market, fully hybridized car in the world; yet at the same time, its former chief Katsuaki Watanabe (in office 2005–09) had been bringing gas-guzzling SUVs like the Highlander onto the market and using them to attack the last bastion of the American Big Three. These vehicles, wildly successful in some regions of the USA, meant big turnover for all companies manufacturing them – right up to the outbreak of the crisis. Then suddenly, Toyota and its American rivals were left sitting on them; and despite what is generally considered to be the very flexible Toyota production system, the carmaker kept churning SUVs out, filling the storage facilities of dealers who were having trouble selling these vehicles. Toyota had to radically scale back production, sometimes resorting to shutting American and Japanese factories for months. Then, just as demand was recovering in the US market and Toyota was planning to ramp up in-country production, the earthquake shook its home base and sent the company reeling. In mid-2009, Toyota was producing too many cars for the American market: in mid-2011, Toyota literally could not produce enough. Toyota has acted fast to overcome the effects of the earthquake and subsequent Fukushima catastrophe and reported in its 2011 half-year figures that it will return to fully flexible, pre-quake planned production by the end of the year, both at home and abroad.

Production abroad, heavily disrupted by the interruptions to supply chains following the catastrophe of March 11, has become more important to Toyota than ever in recent years. Throughout late 2010, Toyota had been stepping up its production of SUVs such as the Highlander, including a US$500 million (€347 million) investment in its Princeton, Indiana assembly plant. The move was intended to help shield Toyota from fluctuating exchange rates and increase its North American production to more than 60 percent of its US and Canadian sales. The new Mississippi factory, too, which had previously been put on hold, was scheduled to build the Corolla compact sedan beginning in the fall of 2011: and provided that Toyota does return to full production soon (as most observers expect it will), starting in Mississippi, nearly all Corollas for the North American market will be built there. Some Corolla production had been transferred back to Japan following the closure of the Nummi plant

(Fremont, CA) in April 2010, but Toyota is once again trying hard to up its production abroad.

There is a good reason for this. In addition to woes of its own making – such as the recalls crisis – Toyota has a Sword of Damocles swinging over its head: the yen. The Japanese currency has been getting stronger and stronger since 2008 (the dollar has been getting weaker and weaker), and this has a significant effect on the export of Toyota's products. This trend has accelerated rapidly in 2011 and shows no signs of slowing, and this sudden gain in the value of its own currency makes Toyota cars more expensive for customers abroad, above all in the USA, where the dollar has been depreciating significantly since the outbreak of the economic crisis. A strong yen against a weak dollar, in particular, has several simultaneous effects on Toyota's bottom line. On the one hand, the price of raw materials tends to decrease, since these are usually paid for in dollars; on the other hand, this is an advantage that Toyota's international rivals enjoy as well. Nevertheless, Toyota's dollar debts also shrink when the dollar loses value, and this, too, is a boon for the balance sheet. Yet this is not enough to balance out the negative effects of upwards pressure on the yen, due to the large percentage of exports in Toyota's sales. Japanese carmakers overall put a figure on these negative effects: every 1-yen gain by the Japanese currency against the dollar reduces annual operating profit by ¥16 billion ($187 million, €129 million) at Honda and by ¥30 billion ($322 million, €224 million) at Toyota, according to those companies.

In fact, when the surging yen was at a 15-year high compared to the US dollar in August 2010, Toyota President Akio Toyoda caused something of an uproar in the automaker's home market of Japan when he commented that "logically, it doesn't make sense to manufacture in Japan." Of course, in the aftermath of the earthquake, there was suddenly no place for this kind of talk: despite the even more drastic appreciation of the Yen since March 11, Toyota is part of Japan Inc. and shows no signs of shirking the responsibility this entails in times of national crisis. Speaking to CNN on a visit to Seattle in April 2011, Akio Toyoda went on record as saying that the 3 million vehicles the company currently produces in Japan would still be produced there. What he failed to comment on was the extent to which the figure of 4 million vehicles already produced abroad may rise in future as any further production increases occur away from the home islands.

Nevertheless, due to the slow US recovery and the weak market in Japan, physical increases in capacity are currently not top of the agenda; there will be some new facilities in the earthquake region and new capacity in emerging markets to avoid both import tariffs and the effects of the yen – the former follows a political and emotional rationale and will be unprofitable whilst the latter, conversely, follows from exclusively economic reasoning. Since the last calendar quarter of 2009, Toyota has been adhering to a strict emergency savings program that is intended to lessen the high costs that result from unused plant capacity. In the past, Toyota had consciously accepted the high costs of depreciating its fixed capital as necessary evil consequences of its expansion. Yet the crisis revealed the high level of capital expenditure as a weakness in the Toyota expansion strategy, which resulted in the carmaking giant putting itself on a draconian course of savings that required a reduction of capital expenditure by JP¥650 billion (US$6.9 billion, €4.8 billion) by the 2010 financial year, which would essentially have meant a halving of the previous level of capital expenditure. As seasoned observers of Toyota might have expected, the carmaker exceeded its own targets and reduced its capital expenditure to JP¥579 billion (US$6.2 billion, €4.3 billion) in the 2010 fiscal year – from a starting figure of JP¥1.3 trillion yen (US$14 billion, €9.7 billion) the year before. Toyota expects investments in property, plant, and equipment to be approx JP¥620 billion during fiscal 2011; investments include about JP¥400 billion ($5.4 billion, €3.7 billion) in Japan and more than JP¥100 billion ($1.1 billion, €800 million) in North America.

Spending on research and development was also expected to drop from JP¥958.8 billion ($10.3 billion, €7.2 billion) in the 2008 fiscal year to no more than JP¥820 billion (US$8.8 billion, €6.1 billion) in 2010; yet here, too, Toyota outdid itself and spent just JP¥725.3 billion (US$7.79 billion, €5.4 billion) on R&D, which is less than it spent in 2006 and only 3.8 percent of consolidated net revenues. Despite this brutal cut, Toyota is still the automaker that spends the most on developing alternative engines. Toyota's forecast for total R&D spend in fiscal 2011 is stable at JP¥740 billion.

The company, which likes to boast that it can wring water out of a dry towel, wanted to prove to itself that with this extreme savings program it had the ability to lower costs without having to lower its performance; and indeed Toyota did manage to meet and exceed its savings estimates, which helped it to get back in the black in fiscal 2010. This speedy return to operational profitability is a success story of a very Toyota kind, but the humiliation of the recalls has soured the story somewhat in the eyes of the world. Above all, it was the worldwide and very public discussion of Toyota's growth strategy which, as humbling as it was, led to a change in the growth strategy it had otherwise followed for decades. The official and unmistakably rueful tone of Akio Toyoda's statement before Congress marked the turning point from a Toyota giant feared across the world for its gargantuan sales figures to an organization that has forced itself to shrink. There can be no more talk of the Japanese company aiming for volume above all: "the World number one does not always have to be the biggest, just the best," said Akio Toyoda at the Nürburgring 24-Hour Race in Germany in May 2010.

After all, Toyota's drive to size created several weak points. Simply put, Toyota had been overstretched after expanding production and marketing at a previously unseen speed, an expansion that was still in full motion even as the financial crisis broke out and scared consumers into shutting their wallets. The giant was showing signs of fatigue, as evidenced by recalls even prior to the sticking gas pedals and "unintended acceleration" disaster in 2009–10. Several hundred thousand vehicles had already had to be called back between 2005 and 2008 to have various defects repaired, and this alone was an incredible slip-up for a carmaker which considers itself fanatical about quality: "We will only put products that have been sufficiently checked to the judgment of the world" is another of the guiding principles which Sakichi Toyoda, one of Toyota's "fathers", imparted to the company from its earliest days.

Yet the breakneck speed with which Toyota was forced to lengthen its purchasing and logistics supply chains put the checks and balances in the carmaker's processes under too much strain, and the proud company could no longer live up to its own expectations. Suppliers, too, were struggling to keep up with the speed of the march, and although the man at the top during the expansion, Watanabe, was considered an expert in the fields of production and purchasing, Toyota became increasingly cumbersome under his direction. Furthermore, certain Japanese cultural necessities, such as always making decisions as a participative group after consensus has been reached, may well have contributed to increasingly sluggish reactions by Toyota. As it is, extreme prudence in reaching decisions is a part of Toyota's genetic code, as experienced observer Masaaki Sato writes: "Toyota's decision-making, in general, was so excessively cautious that the company was ridiculed for knocking on a stone bridge before crossing it."

This ponderous manner remained even in the face of the first signs of encroaching quality assurance issues. In 2006 alone, over half a million cars had to be called back, but former CEO Watanabe went on record to contradict rumors about those recalls being related to an overly speedy pace of expansion. Yet even at this early stage, the atmosphere within Toyota was already tense. According to the *Harvard Business Review*, Toyota managers were "gravely concerned about not having enough people to sustain their global growth. In fact, almost every aspect of Toyota is straining to keep pace with the company's rapid expansion and with technological change,"[3] and Watanabe was not able to send Japanese managers from Toyota City out to new sites abroad at anything like the pace he wanted. The idea was to send supervisors – termed *sensei* – to make sure that the strict terms of the Toyota production philosophy were being respected as strongly in factories abroad as at home; but a year later, in 2007, Watanabe complained that he had only been able to send 2000 *sensei* to Toyota's production facilities, and that three times that number were already required. At the same time, Toyota opened training centers in locations such as the USA, Thailand, and Great Britain in order to increase the integration of employees and suppliers abroad into the

"Toyota Way," which had long become a worldwide synonym for efficient, high-quality manufacturing. The peculiarities and standards of this production philosophy, known as the Toyota Production System (or TPS) can only be implemented abroad after exceptionally intense workforce training, and this training had fallen far behind the speed of the expansion abroad. The results of this heedless dash ahead caught up with Toyota, and quality assurance slips became the most serious threat to the carmaker. Even now, after an earthquake, a tsunami and a nuclear disaster in its home territory, it is still the quality issues that cast the longer shadow over Toyota's balance sheet in its most important foreign market, the USA. Natural disasters, terrifying as they are, can be surmounted – much of the work done in getting production back to normal is a matter of creative logistics and hard work – but a seriously damaged reputation is far harder to repair.

Back in 2007, when Volkswagen publicly declared that it was trying to overtake the Japanese company, almost nobody believed this was possible. Now, just four years later, the defending champion is weakened and its contender has been able to profit from two years of good growth in the automotive market: 2011 may well be the year in which Volkswagen does indeed overtake Toyota. "We will continue to gain market share in the crisis. As a Group composed of many brands, we not only have the size, but also the economic and ecological potential to do so – more so than any other car maker," said Volkswagen CEO Martin Winterkorn in 2009; two years later, and VW's market share is indeed up. What was new was the additional stated intention to be a leader in environmentally responsible technology, and of all possible fields it could have chosen, Volkswagen now wants to fight a battle on the ecological innovation front: Toyota's core area of strength.

3.2 A NEW BEGINNING

Trying to get a grip on the crisis, Toyota put many hopes on its new and relatively young chief Akio Toyoda, who came into office somewhat earlier than expected. The grandson of the great founder had the job of steering Toyota out of its biggest-ever loss and breathing new life into the much cited Toyota narrative, with its roots in postwar Japan – based on the virtues of solidarity, loyalty, efficiency, discipline, modesty, and austerity. Not surprisingly, the last three virtues were held by many to have died during the expansion years. In Japan, Toyoda was expected to clean up the company, cut *muda* (a Toyota term for waste – see Section 3.4) and help the giant get back on its feet and ready to take on the other manufacturers again.

Just one year after taking on the job, Toyoda made his mark on the company under the strict savings program: in August 2009, Toyota had to announce the closure of the Nummi joint venture with General Motors in Fremont, following GM's disclosure that, in the wake of its filing for Chapter 11, it would be forced to stop producing the Pontiac there. Production of the Corolla and the Tacoma at the Fremont plant had been shifted partly to other sites in the USA and partly back to Japan. Nevertheless, in contrast to GM, Toyota offered the Nummi staff, many of whom were unionized in the UAW (United Auto Workers), a severance package totaling US$280 million (€195 million).

Out of the race – Toyota quits Formula 1

The crisis forced Toyota to take further, radical cost-cutting measures. In November 2009, the car manufacturer announced that it would not continue its involvement in Formula 1 racing, a decision which cost 500 jobs in Cologne, Germany. Toyota had already abandoned its Japanese racing circuit to save costs, despite the fact that it had acquired Fuji International Speedway – one of Japan's most important race tracks – only eight years earlier and that renovation costs there had run to JP¥20 billion (US$215 million, €149 million). It was not ready for the Grand Prix until 2007, and now the track has closed.

Once again, this was a cut that was clearly difficult for the Japanese – just how difficult became clear during the press conference on November 5, 2009 at which the exit from Formula 1 was announced, where Tadashi Yamashina, head of Toyota's motor sports department, broke down in tears. Later, Toyoda himself followed the announcement by saying that the decision to withdraw from Formula 1 activities had been one of the most difficult since taking office, especially in view of the job losses it had caused.

Along with this savings program, the heir to Toyota also cut some important tenets out of the company philosophy: "We are in a phase of difficulty, and the whole company is looking for a savior. Let me be clear: I am not this savior," said this newly crowned king at a speech in Tokyo in October 2009. As a member of the founding family, these words were an unexpected break with Toyota ideology.

However, Toyoda had already spoken with unusual and brutal candor (for Japanese tastes) about Toyota's dip into the red, publicly admitting the loss of face: "We have stretched ourselves too far," said the chief, pointing to company ideology as a second source of the disaster aside from the sales crisis. "Toyota is but one step away from capitulation to irrelevance or death," he said at the beginning of October 2009 – a very dramatic warning that he delivered to his employees. The colorful expression was borrowed from management icon Jim Collins who, in his book *How the Mighty Fall*, describes the five stages of corporate decline. According to Toyoda, the first stage, "hubris born of success," came when the organization became arrogant in the wake of its global success, moving into the second stage which Collins terms "the undisciplined pursuit of more," followed by the third stage, the "denial of risk and peril," and the fourth, termed "grasping for salvation." After that, there is only "capitulation to irrelevance or death." Toyoda used this terrifying prospect as a backdrop against which to ask his managers and workers to start thinking differently. Indeed, though this depiction of decline seemed somewhat exaggerated at the time, it quickly became a reality in the months that followed.

After the quality crisis, Toyota had to reinvent itself – and its young CEO accepted the challenge. Unveiling the new "Global Vision" in March 2011, Akio Toyoda outlined Toyota's new mission statement as well as profitability and regional targets, aimed at marking a new chapter for the world's largest automaker. "I began to articulate for myself what kind of company I want Toyota to be", Toyoda stated in introducing his presentation about the carmaker's core values and mission. Apart from the traditional principles of the Toyota Way and the Toyota Precepts ("always making better cars, enriching communities' lives") Toyoda identified emerging markets and environmentally friendly cars as the key pillars of the new strategy. Toyoda said the carmaker would launch ten new petrol-electric hybrid vehicles by 2015.

> As for our series of quality problems, I, personally, played a direct role in explaining to the US Congress, to members of the media around the word, and to, most of all, our customers, what kind of company we are and what kind of values we hold. That experience led me to keenly feel anew the importance of making it known to all our customers and to the public at large the direction our company intends to take. Together with my personal reflection on the matter, it awakened me to the need for articulating our vision for Toyota.
>
> Akio Toyoda, March 9, 2011

Though it is late, it is not surprising that China is to play a key role in the new strategy: in the People's Republic, Toyota has been lagging behind competitors such as VW. "At Toyota, we hold especially high hopes for the Chinese market," Toyoda announced in his remarks on the global vision. He set a new target of securing 15 percent of the carmaker's global unit sales in China as soon as possible. He also declared Toyota would develop a full range of plug-in hybrid vehicles, pure electric vehicles, and fuel-cell vehicles. "Our idea is to position Toyota to be ready with whatever technology captures the imagination of the car-buying public," he concluded his speech.

In 2010, emerging markets made up 40 percent of Toyota's sales – by 2015, Toyota wants to achieve a geographical balance in its sales portfolio between industrialized nations and emerging markets. In terms of profitability, Toyota aims to achieve an operating profit of about JP¥1 trillion "as soon as possible" – that would be nearly double the JP¥550 billion operating profit projected for the 2011 fiscal year.

Toyoda also wants to lift the annual operating return on sales to 5 percent from an estimated 2.9 percent for the current fiscal year. He said the automaker will raise profitability to achieve its goals even in severe situations where the dollar is weak at JP¥85 and the company's sales volume only 7.5 million units. "This is the bottom line for a sustainable growth as we aim to build a structure to generate profits when our sales fall by about 20 percent in case we see another major economic downturn," the CEO said. Furthermore, the automaker will reduce the number of its board directors to 11 from the current 27 in an effort to speed up decision making, after Toyota last year announced steps to reshape global operations to improve the recall process.

Yet Toyota would not be a distinctly Japanese company if it did not flank its new beginning with lessons drawn from its history and its founding mythology. Therefore, before we ask whether its corporate philosophy and culture can help steer the defending champion back to its former strength, we must take a closer look at Toyota's origins, at the cradle from which this giant, long feared and admired in equal measure, began to grow.

3.3 THE POWER OF TRADITION: HOW A JAPANESE LOOM BUILDER BECAME AN AUTOMOTIVE GIANT

Toyota has always been proud and respectful of its history, and most observers would agree that the automotive champion indeed has a lot to be proud of. In developing its production from scratch, step by step, Toyota also developed the legendary Toyota Way, a production system that has become known and admired in all sectors of industry the world over.

The development of the Toyota Way began while Toyota's parent company – today named Toyota Industries – was still making looms. Japanese inventor Sakichi Toyoda set the ball rolling in 1924 when he, together with his son Kiichiro, constructed an automatic weaving machine (the "Toyoda Automatic Loom"). It was here that the first principle of the Toyota Way was employed: *Jidōka,* also known as autonomous automation. *Jidōka* means that the machine stops itself whenever there is a problem, which was a truly revolutionary property at the time. Sakichi developed a mechanism that stopped the weaving machine whenever a fiber tore, and this instantly put Toyota in a class of its own in terms of efficiency, because at Toyoda Automatic Looms a far smaller number of people were able to work a much greater number of machines. Prior to this innovation, each machine required a worker standing nearby to watch for torn fibers. Today, *jidōka* remains one of the most important production principles at Toyota, and beyond that has a special place in Japanese industrial history. This early innovation in the cradle of the carmaker was already pointing in the direction in which the adolescent giant would be marching on its way to the top: the efficiency of production and of the factory as a whole, as well as the improvement of single steps in manufacturing processes, have since those early days been of primary importance to the organization. *Jidōka* is what brought Toyoda Automatic Looms success and enabled it to found the Toyota Motor Corporation.

It was also Sakichi Toyoda who introduced the "five whys" as a problem-solving instrument, a philosophy which was picked up, adapted, and eventually made into a standard management strategy across many manufacturing industries. Whenever there was a problem, Sakichi asked "why" five times in order to analyze the cause at the bottom of the issue. He would then change the production process accordingly, until the problem would disappear. This is another principle which is still employed at Toyota more than 70 years later.

Toyoda's revolutionary and cost-saving production processes eventually allowed Sakichi to sell the patent for the automatic loom for enough money to finance the construction of an automobile production plant as the car industry started to gain in importance in the 1930s. On August 28, 1937, this branch of Toyoda Automatic Looms became the independent Toyota Motor Corporation. Toyoda family members continued to exercise a decisive influence on the Toyota production system. Kiichiro Toyoda studied the production technology used by the Ford Motor Company in Detroit and, upon returning to Japan, transferred the Ford system to smaller production volumes, cutting down on waste despite the smaller numbers of vehicles manufactured in Toyota City. Wasting resources is to this day one of the capital sins at Toyota. The organization is on a continuous crusade against waste of all kinds – which are defined as overproduction, superfluous transportation, waiting time, and correcting defects. Avoiding these forms of waste is one of the pillars of Toyota's world-beating efficiency.

What is muda *at Toyota?*

> Take the movement of parts in a factory, for example. Moving components doesn't add to their value; on the contrary, it destroys value, because parts may be dropped or scratched. So the movement of components should be limited as much as possible. I want our production engineers to take up the challenge of ensuring that things move as little as possible – close to the theoretical limit of zero – on shop floors.
>
> —Former CEO of Toyota, Katsuaki Watanabe, 2007

When Eiji Toyoda took the helm in the 1960s, he, too, went to the United States to study the American car industry intensively. As soon as he returned to Japan, he started restructuring the Toyota Motor Corporation, with one of the decisive changes he undertook being what was known as the "Ford Motor Company suggestion system," the concept founded on the notion that every employee should help contribute to improving work processes by making suggestions to their line managers. This early form of the employee idea system has been part of the Toyota Way ever since and, ironically enough, was brought back to Western car companies (who had lost sight of it) by the spread of the "Toyota Way" in the 1990s. At Toyota, the introduction of the Ford suggestion scheme laid the foundation for the continuous improvement that it developed further into the legendary *kaizen* system.

> Manufacturing is the foundation of civilization.
>
> —Eiji Toyoda

The Toyota Production System is in continuous development, and as such was fundamentally shaped by Taiichi Ohno, an acknowledged master of production processes. It was he who introduced just-in-time manufacturing to the company, now an industry standard: its essential aim is to synchronize consumption and production, saving unnecessary storage and thus cutting down on *muda*.

Toyota adopted the just-in-time system in the 1950s and 1960s, gradually rolling it out over the entire production system and extending it to suppliers. While automotive manufacturers in Europe and the USA were still piling their components high, Toyota's suppliers delivered their components directly to the assembly line at just the moment they were needed. For Toyota, this was a clear advantage over its main rival of that time, Nissan, which was then producing in higher volumes than Toyota and was thus in a better position to make savings. It was the need to catch up with competition that drove Ohno to keep improving the production process, which led to another invention, the "pull system," which turned the existing organizational processes literally on their head. The pull system is the cornerstone of just-in-time manufacturing and the whole concept is based on customer demand. This demand is known as "pull," and runs backward. The dealer became part of the production system as Toyota gradually stopped building cars in advance for unknown buyers and converted to a building-to-order system in which the dealer was the first step in the system, sending orders for cars already sold to the factory for delivery to specific customers. The breakthrough was the extension of the *kanban* (signboard – see below) system to the dealer; the actual number of cars at the dealership was the "pull" signal for the factory, and replaced a planning process in management committees.

As soon as demand is established in the form of a confirmed order, the organization begins production by sending signals in a backward direction. Instead of "pushing" units from one workstation to the next – i.e. from the press shop to the welding and paint shop, Toyota started "pulling" its units through the system. Consequently, all workstations send their orders upstream: beginning with the particular order for a particular car, information about the quantity of stamped parts or components needed, for example, is passed upstream – i.e. from body shop to stamp shop and back to Toyota's suppliers. The secret of this system is that it automatically passes the information or order right to the starting point of production, which means that, at Toyota, only so much was produced as could be passed on to the next workstation, allowing Ohno to reduce the material inventory buffer to a minimum. In order to direct this process, he used a signboard (*kanban*) to pass orders back down the line and installed signals (*andon*) so that each worker could stop the assembly line if a problem arose.

Part of the Toyota philosophy is that a team leader who sees an *andon* signal intervenes immediately in order to solve the problem with the employee directly, and employees are expected to try to recognize the cause of the irregularity themselves and, if possible, propose a solution. This form of autonomous working allowed workers to stop the production process whenever there was a disturbance to it or an irregularity, the elimination of malfunctions being for Ohno "built-in quality." Not only was he successful in smoothing out processes, but also in reducing the considerable quality issues which had plagued Toyota in this early period. Nevertheless, it would be years before these methods spread from Toyota City to the West.

At the start, however, the Ohno production system was by no means near perfect. The first attempts at using the revolutionary pull system were made in 1948 in a forgotten corner of the factory; yet from the very beginning, Toyota was only able to succeed in the Japanese market by concentrating on high quality, low costs, and low prices through a relatively small volume. The historical background for this is that, in the early part of the twentieth century, Japan was still a predominantly agricultural society, meaning low wages and negligible domestic demand; for this reason, the car market in Japan developed in small steps and required low-priced models. These constraints made Toyota a manufacturer that, more than any other, has always kept on improving its production system.

Back then, of course, continuous improvement was less a question of philosophy than of survival. The newcomer Toyota had to base its expansion – and its profits – on taking market share from established competitors, i.e. it was faced with challenging benchmarks in terms of prices and quality, and the Toyota Production System (TPS) was the answer to this challenge. Later, TPS became Toyota's competitive edge, increasing its profits and giving it a decisive advantage, first over its Japanese rivals and decades later over its formerly superior Western competitors. The will to continually improve became part of the ethos of the organization itself; it is an essential gene of the industry leader as we know it today. Nowadays, Toyota places emphasis on consistent, continual, structural adaptations to the assembly process, quality control, and the supply of materials.

After Ohno's first attempts to roll out the new production system, it took 14 years to implement it in all Toyota plants, even though no one outside of Toyota realized this. As author Steven Spear remarks, "since the 1960s, Toyota has been more productive than its competitors,"[4] and yet this fact remained unnoticed long after Toyota had started its conquest of automotive markets worldwide. The deliberate and cautious improvement that eventually led Toyota to the industry's top long passed unremarked by the company's competitors, which slept on until their Japanese rival was far ahead.

Indeed, it was not until the 1980s that the TPS slowly started to spread through the USA thanks to the joint venture with General Motors, Nummi. This gave Americans the chance to experience the system first-hand at the factory in Fremont, California, which under Toyota's management was soon to become the most productive automotive plant in the country.

MIT researches "Toyotism"

"The Machine that Changed the World" was the title of a study produced by the Massachusetts Institute of Technology (MIT) on Toyota's production system in 1991.[5]

In this study, the authors termed Toyota's methods "Toyotism" and held them up as the next stage in industrial history after "Fordism." Analyzing the TPS, the authors explained why they saw Toyotism as a successor to Fordism, building on Fordism's structures but eliminating its weaknesses. This study was the first to use the term "lean production" for the TPS, which it contrasted with "mass production" of the Fordian type.

According to the study, lean manufacturers draw their advantage from having learned how to cover costs while producing in comparatively small volumes. The authors showed how Toyota saved costs while pressing and stamping metal by producing smaller batches, thus reducing transportation costs for the heavy parts and saving on expensive warehousing. Storage and stocking, however, were major components of any mass-production system.

Furthermore, the small batches Toyota produced at the time had the advantage of making any defects visible earlier on, and this in turn had wide-reaching effects on staff, who automatically became concerned with quality assurance. Waste due to massive quantities of defect parts could thus be eliminated. Nevertheless, this kind of system is highly dependent on workers, who need extensive training and high motivation. If individuals did not spot problems early on and find solutions on their own, the assembly line had to be stopped by using the *andon* chord.

Traditional mass producers, on the other hand, were forced to accept mistakes that had been noticed on the assembly line and correct those after the vehicle had been finished, or to take the vehicles out of the production process and repair them later. Both strategies required specialist workers and incurred high costs; these costs, however, were considered by mass producers to be marginal compared to those involved in stopping the assembly line. Toyota, on the other hand, clearly considered this to be waste, because none of the specialist workers actually added value to the car. The Japanese, in contrast, reasoned that the mass-production practice of passing on errors to keep the line running caused errors to multiply endlessly.

Instead, Toyota Production workers were taught to trace every error systematically back to its ultimate cause, then devise a solution, so that it would never occur again. This required a highly specific division of labor and permanent quality control, but allowed any defects to be rectified speedily.

Not surprisingly, as Toyota began to experiment with these ideas, the production line stopped all the time. However, as the work teams gained experience identifying and tracing problems to their ultimate cause, the number of errors began to drop dramatically. Today, in Toyota plants where every worker can stop the line, yields approach 100 percent. The line practically never stops.

Toyotism relies on a high level of autonomy for its teams of workers and on a system of incentives for staff, including safe, high-paying jobs with a good chance of promotion, which are strongly linked to group effort. However, the study also cited the extremely high intensity of the work.

Starting at Nummi, the TPS and lean manufacturing methods conquered almost every manufacturing industry sector in one way or another. Nowadays, they represent nothing less than the central tenets of modern production, and not just in car manufacturing.

3.4 THE LEGEND AND THE REALITY BEHIND IT: THE TOYOTA PRODUCTION SYSTEM

For more than three decades, Toyota's system has been one of the most important industry benchmarks for high-volume car production, and the practice has ushered an entire area of the consulting industry into existence.

There was no small amount of jealousy in the automotive sector when competitors noticed Toyota's continually high profitability and ever-increasing sales volume. Managers from car manufacturers, large and small, went to Japan by the busload to see what lessons they could learn on how to produce in a lean manner – i.e. with as few resources as possible and by eliminating waste. The explanations for Toyota's competitive edge that resulted from these production pilgrimages are as numerous as the managers who went on them, and this in itself is proof of just how often other companies have tried – and failed – to analyze or implement the TPS in their own factories.

> There is no end to the process of learning about the Toyota Way. I don't think I have a complete understanding even today, and I have worked for the company for 43 years.
>
> Watanabe[6]

Since its "discovery" at the Californian joint venture with GM, a number of elements of the TPS have been implemented in automotive plants across the world, and manufacturers and suppliers everywhere in the industry have adopted methods like "just in time" and "just in sequence" as a basis for general competitiveness. Moreover, the basic concepts of the TPS have long been refined and evolved into more modern production methods – in Toyota City as well as in the rest of the industrial world. But even now, as Toyota terminology like *kaizen*, for example, has become standard vocabulary among industrial managers, the actual core of the Toyota Way never really made it into the factories of the West. Most manufacturers have settled for the explanation that the Toyota Way is a combination of single factors, and falsely hoped that by implementing some or most of them they could reach the high level of efficiency at which Toyota has set the bar.

Yet the reason why the Toyota Way is more than just the sum of single instruments, more than just continual improvement along with cutting down on waste in the value-creation chain – and, most importantly, why it does not automatically work outside of Japan, even in Toyota-run factories abroad – will be explained in the next chapters. We will take a close look at two aspects: the Toyota philosophy and its meaning for the way the TPS works, and the sociocultural factors that are at work in the Japanese manufacturing environment and allow it to slim down its factories to their much-envied level of efficiency.

Kaizen

In the West, *kaizen* is generally understood as a call on employees to make suggestions to plant management about how processes can be improved. This was the form in which it arrived in the 1990s, not only in almost every car factory outside Japan, but in many production sites all over manufacturing industry. One of the most prominent adopters in Germany was Volkswagen, which – like many German companies – translated *kaizen* into *kontinuierliche Verbesserung* (German for "continuous improvement") and made it into a program named KV2. TPS experts from Japan were shipped in to oversee its implementation, and under KV2, several of the principles of TPS were put into place, which are still in practice in Wolfsburg today. One of these, for example, is the visualization of processes and movements in the plant.

Visualization

This principle has since spread across manufacturing industries and means the display of information related to materials and tools in signs that are understandable by every plant worker, irrespective of language or background. An example of visualization might be tools being hung on a wooden block when they are not being used, rather than lying around at a workstation. Each tool has a designated spot on the block, and this is indicated by the silhouette of the instrument in question. This makes it easier for workers to see whether tools are missing and allows staff to keep better control of plant equipment.

This method was of great use to Toyota, which continually ran up against language problems during its speedy global expansion: not only did overseas workers not speak Japanese, but the Japanese did not always have a complete grasp of the language of the country in which they were producing. Toyota therefore overcame language barriers and standardized production across the globe by displaying as many as possible of its processes in pictures. The next stage of visualization is the *kanban* system, of course, which is used to pass information between workstations and offers a simple but comprehensive information display that makes it possible for everyone in the plant to respond quickly to problems and to understand the plant's overall situation.

Undoubtedly KV2 has led to improvements at Wolfsburg, and yet it remains unclear whether the system has had a sustainable effect on productivity. The essential problem with TPS in all industrialized Western societies remains the same: once one set of TPS-style measures (e.g.

employee suggestion or visualization) has been implemented, a fundamental root-and-branch reform of the production culture into a "learning organization" is pushed to the bottom of the agenda, or forgotten completely.

Yet *kaizen,* as Toyota understands it, is more than a combination of measures. In the TPS, *kaizen* is the central cost-control instrument and is crucial to the entire production organization right down to employee performance measurement. The culture of permanent cost reduction and the concurrent increase of productivity at Toyota cannot realistically be explained only by the tools applied the improvements undertaken by employees of their own free will. Rather, it is far more a question of goals set by management and consequently accomplished by the workers. *Kaizen* as a combination of strict cost discipline and continuous optimization of process efficiency is deeply anchored in all Toyota employees. It characterizes teamwork on the company's assembly line, the social rules among the employees, and the calculation of their salary. In short, it is part of Toyota's sociocultural fabric, its culture and therefore part of its anatomy; it is the spine of this giant that stands so tall in the face of its competitors.

The kaizen *philosophy*

At first glance, *kaizen* means nothing more than continuous effort with the aim of improving; it is the basic attitude of not being satisfied with what has already been achieved. Toyota has not just systematized this attitude, but has institutionalized it, too. The mantra behind *kaizen* is that every product and every process can always be improved, and that this improvement should not be left to chance – rather, it should be built into the system. Toyota has codified a process that starts with planning, proceeds through implementation, verification, and adaptation and goes back to planning, forming a closed circle. In Toyota terminology this is referred to as "plan, do, check, act, and plan." This system forms hypotheses about processes, applies them in practice, and tests them, either discarding or implementing them – before planning on the basis of what has just been achieved. It is a continuous, never-ending series of attempts to reach the optimum.

Kaizen is also, however, a collective term for a whole package of routines that workers have to carry out, and workers are very highly rewarded (not just financially) if they achieve the *kaizen* goals set for them by management and manage to improve their work through suggestions or accomplishing the *kaizen* quality circle.

Just how Toyota uses *kaizen* and exactly which prerequisites are necessary for these principles to have their desired effects only becomes clear with a close look at the Japanese shop floor.[7] In Japanese Toyota plants, a far higher number of tasks need to be delegated by team leaders than in a traditional Western production environment. In addition to core production activities and secondary line work – i.e. quality assurance or scheduling material delivery – there is a whole series of other tasks, such as training the so-called "freshmen." Freshmen still tend to be trained on the job rather than beforehand, carrying out *kaizen* routines or taking care of the social and interpersonal aspects of work life. This means that *kaizen* is deeply embedded in team dynamics, and people who sign an employment contract with Toyota do not just begin a new period in their professional lives (as we put it in the West), but begin their company life. There is far more to the oft-cited Toyota culture than just a particular way of building cars; just a look at the qualifications of Toyota employees and the measures taken to integrate them into the company is enough to show why it has sometimes had trouble implementing its own rigid corporate culture in other regions with heterogeneous cultural backgrounds, such as Europe.

In the Japanese core production facilities, the freshmen – usually straight out of school or university – are assigned to a team in which they remain for a long period. This team is a nursery ground to the extent that it socializes the freshmen and teaches them the basics. Teams are close-knit, and the freshmen get all the help they need to perform the task at hand; in return, it is expected that new team members show strict loyalty and prove that they are fully committed to Toyota culture and the company, which might include voluntary overtime, for example. The TPS relies on teams that are close and offer solidarity to all members. Only with functional teamwork can problems in daily production be solved effectively and in a short amount of time. As training continues, freshmen are slowly allowed to grow into leading roles within the team, including the possibility of formalizing their advance through the ranks by being appointed team leader, although naturally not every freshman is likely to complete this path.

The team leaders have a central function in *kaizen:* they are responsible for passing on the tenets of the Toyota Way and seeing that these are adhered to – just-in-time production, *jidōka,* standardized tasks and time targets, *kanban,* correct use of the *andon* line, avoiding *muda,* calculating production efficiency, and closing the *kaizen* circle. In contrast to Western business culture, team leaders do not just pass on knowledge, but cultivate behavioral codes in their teams: Toyota people learn to work and act as expected by the TPS, and a cornerstone of this mindset is that productivity must increase continually – even though in incremental steps. If productivity cannot be improved as expected, the entire team loses face; if it hits or even exceeds targets, Toyota rewards the entire group, just as it does for successful suggestions regarding improvements.

The actual principles behind TPS were not codified until recently. The system functioned for years – and still functions – by being passed on to new employees, socializing them from their first work day onward. That's how it has become rooted in the subconscious of the collective and of each team member. In addition to this, top management gives savings targets that are broken down to the level of production costs per individual vehicle. When setting these targets, Toyota consistently works on the premise that what is considered "standard" today will no longer be sufficient tomorrow and so must be improved.

The efficiency with which Toyota implements and sometimes even exceeds savings goals became clear in the Emergency Profit Improvement Program that brought the industry giant out of the red in record time. At the start of the program in November 2008, the *kaizen* goal for cost reduction in the 2009 fiscal year was set at JP¥130 billion (US$1.4 billion, €1 billion) and at approximately JP¥800 billion (US$8.6 billion, €6 billion) for fiscal 2009 and 2010. Of the total target, JP¥430 billion (US$4.6 billion, €3.2 billion) was to come from fixed and JP¥340 billion (US$3.6 billion, €2.5 billion) from variable costs. In every calendar quarter of 2009 (2010 in Toyota financial years), Toyota more than met its own goals for savings and set the bar higher every quarter. Over the entire period in question, Toyota did not just manage to push its costs down by JP¥800 billion, but by double this figure: it saved an astonishing JP¥1.690 trillion (US$18 billion, €12.5 billion). The same happened with Toyota's predicted sales and its losses – "loss" was finally crossed out in favor of "profit" in spring 2010: it had become an operating profit by the time the annual figures were published. At the time of writing, it looks very much as if 2011 will provide another prime example of how Toyota consistently overperforms on targets it has set for itself. In the immediate aftermath of the March 11 earthquake, Toyota estimated that it would need eight months to restore production; by the time it released half-year figures in late July, it was able to report 90% production in June and forecast "near pre-earthquake planned levels in July." Moreover, the company reported that it would "return to full production able to flexibly meet consumer needs in November and December", which analysts interpret as Toyota for: the manufacturer will utilize its production facilities to their absolute maximum, and produce as many additional vehicles as possible in an attempt to reclaim market share where it was lost in the aftermath of the earthquake.

This dynamism comes from what lies at the heart of the Toyota philosophy: whatever the current target is, it must always be exceeded, and the next must then be increased correspondingly and then also exceeded, or at least reached more quickly, and so on. "We aim to exceed expectations," Toyoda declared when he outlined the Global Vision, Toyota's mission statement for the coming years. The mythical narratives that emanate from Toyota, such as the development of the first Prius, all match the dramaturgy of the "upwards spiral".

Legend has it that the Prius was ready several months before a deadline that engineers initially had considered tight. The realization then dawns that this habit of raising the bar, reaching it, and then raising it further has become an ingrained reflex at every level of the organization. Compared to the culture and habits of Western manufacturers, it becomes clear once again why only the Japanese carmaker from Toyota City could increase its sales by half a million per year and take market share from its competitors in almost every car market on the planet.

Overshooting the targets it sets for itself is just as much a part of the Toyota psychology as *kaizen* or *genchi genbutsu,* and this can explain how Toyota managed to hide the problems resulting from its growth for so many years. Essentially, Toyota's expansion had gradually spiraled out of control, and every time the employees exceeded their goals, they grew more confident and wanted to achieve more: this widespread mindset is one of Toyota's crucial resources, and was what allowed it to make such a Herculean effort in its deepest crisis. Applying the mindset to control and cut costs, Toyota was out of the red within a year.

In fact, *kaizen* philosophy may be a guiding principle at Toyota, but it is the group and team structure living the philosophy that really drive the organization. It is the teams on the shop floor that stage and experience the narrative arc of expecting continuous improvement and achieving even more.

Cost savings targets, the scale on which *kaizen* is most directly measured, are announced to production groups and shop-floor teams and then divided up between areas of work. The responsibility for cost-related *kaizen* lies with the production managers, and cost-control groups are set up at all levels of the organization to check results. This makes the team the central regulatory unit in the TPS: its deeply hierarchical structures (the longer-serving the team member, the higher he/she is ranked) are, however, essentially invisible, and as such the system stands in great contrast to Western human resources management.

The team spirit is further strengthened by Toyota housing, Toyota leisure-time activities, and Toyota City, an entire town in which there is almost nothing that is not linked to the car manufacturer in some way. This interplay of social control and solidarity, this integration of the individual into the collective, as well as the pressure to perform tightly coupled with rewards for doing so, is exactly the right atmosphere to push each employee to reach the highest possible performance. Toyota managers never tire of highlighting the role of the geographical concentration on Toyota City in preserving and passing on the company culture. They describe it as "an excellent environment for nurturing people," a place where the individual is given a sense of security by the collective – and responds to this nurturing by giving his or her increasing productivity back to the business.

> We are in the middle of nowhere; there is nothing to do but work!
>
> —Former CEO Katsuaki Watanabe talking about Toyota City[8]

If workers on the assembly line have reached their *kaizen* target and have been efficient in doing so, their team is immediately promoted in the productivity ratings, which are on a scale from A to D. This promotion goes hand in hand with a higher productivity bonus coefficient that is applied to their basic salary. This means that workers are paid according to productivity, and the profit is shared between them and the company.

Nevertheless, teams that are rated A, for example, have to work hard to stay there: on reaching this top ranking, management will reduce their target time, which often knocks their efficiency rating back down to an average value; and that is not the end of it. One of the most important tools of the TPS is the continuous reduction of personnel at the assembly line, referred to as *shojinka*: this increases output per man-hour and therefore reduces labor costs per vehicle.[9] As far as management is concerned, if a team reaches the A category, this means that is has been given too much time and too many employees, and both are cut. Yet when a team falls back a production category as a result, it will again try to improve its productivity and finally reach the level it had previously maintained – even if this takes years.

In this respect, *kaizen* is in no small part based on the fundamental formula of achieving a higher efficiency in production by reducing staff at workstations, which in turn is achieved by "*kaizen*-izing" the production process for a given vehicle volume in accordance with the scheme of just-in-time production. This form of organized *kaizen,* with the central aim of reducing time taken and personnel allocated per vehicle, is a central motor of the cost-reducing and productivity-growing strategy on which Toyota runs. It allows for an extraordinarily stringent implementation of reforms to costs and organizational structures. The high level of loyalty among permanent full-time employees, the performance-related pay structures,[10] and the lifelong employment guarantee allow for a company culture and organization of labor that could not be more different from structures in the West, and this is what really distinguishes the dynamism and performance of the TPS.

According to recent research, only 10 percent of *kaizen* activities are initiated by the workers themselves, with the other 90 percent coming from team leaders, group leaders, and engineers. Thus, there are two functions of *kaizen* to distinguish: "small-scale" and "large-scale" *kaizen*. Reducing material consumption or the number of employees per task, for example, is large-scale *kaizen* – top-down initiatives emanating from team leaders and managers with the clear aim of increasing productivity and cutting costs. Improving safety at work or rationalizing standard tasks, however, are more small-scale and are often initiated by team members. Together, small- and large-scale *kaizen* are the dynamic core of the TPS. Nevertheless, *kaizen* is mainly a top-down process of managing costs and productivity. Team leaders, engineers, and group leaders have the job of reducing the time taken per task and of optimizing processes, with the overriding aim of reducing staff in their area. This is achieved by hunting down any traces of waste and by keeping all resources – both stocked parts and human resources – down to an absolute minimum.

This means that in Toyota terminology *kaizen* more or less has become synonymous with cost reduction efforts. Even for the less linguistically gifted among us, the quarterly conferences for financial analysts – in which questions are always answered in Japanese and then translated into English – are very revealing, because whenever the topic is reducing costs, the word *kaizen* comes up all too often.

Weaknesses of the Toyota system

"Toyota's way is to measure everything," as Thomas A. Stewart and Anand P. Raman noted 2007 in the *Harvard Business Review*, and it is certainly a fact that no manufacturer in the industry has a better grasp on how to quantify the unforgiving tempo of automotive production and then use the information to its advantage than Toyota.

On the basis of this knowledge, very early on Toyota started to standardize and then streamline its processes. Its factories stand out by being "lean," i.e. they run on the minimum of resources (engineers, workers, machines, and parts are all rationed) needed to reach exactly the production target set for the plant in question. Volkswagen, on the contrary, is carrying weight. The VW method requires a higher supply and inventory of resources, causing higher costs, but its reserves also give it a higher degree of flexibility on the shop floor. A Toyota factory, however, has always been trimmed down to fit an exact production volume and, so long as it hits this ideal level and manufacturing lead times, functions extremely profitably. The plants also work almost entirely defect-free, ignoring for the moment the recall actions, the causes of which probably have less to do with TPS in any case.

> Simple and slim systems make it easier for people to notice abnormalities immediately.
>
> —Watanabe[11]

Most factories that belong wholly to the Toyota Motor Corporation in Japan are relatively old and their architecture imposes limits on just how far modernization can be taken. Nevertheless, the names of these factories – Motomachi, Tsutsumi, Tahara – are industry-wide synonyms for productivity benchmarks, taking top place in study after study by organizations like J.D. Power.

Assembly Plant Awards 2010

In the Asia Pacific region, Toyota Motor Company's Kyushu 2, Japan plant, which produces the Lexus ES, IS and RX, received the Gold Plant Quality Award for Asia Pacific. Out of the North and South American plants, the Toyota Motor Company plant in Cambridge South, Ontario, which produces the Lexus RX 350, took the Gold Plant Quality Award for North/South America.

The Daimler assembly plant in East London, South Africa, received the Platinum Plant Quality Award for producing vehicles yielding the fewest defects and malfunctions worldwide. Averaging 28 PP100, the plant produces the Mercedes-Benz C Class.

> Plant awards are based solely on defects and malfunctions and exclude design-related problems.
>
> —JD Power study, 2010[12]

The rigid, inflexible course that materials and products take through the factory, and through the production network as a whole, creates a stable environment and is an incredibly powerful cost-saving mechanism. This is also true of suppliers, which Toyota likes to keep as close to its production sites as possible, and which adopt the TPS within their own production system – i.e. Toyota suppliers themselves function as leanly as possible.

Yet this constant reflex to save wherever possible undeniably costs Toyota flexibility. If demand suddenly drops unexpectedly, or if the mix of models set out in planning is not selling as expected, the strengths of the TPS – its rigidly constant flow of materials and standardized processes – quickly become a weakness and it loses some of its productivity advantage. This issue became clearer than ever when the financial crisis made inroads into demand and Toyota had real difficulty curbing the rate of production quickly enough across almost 70 plants and scores of suppliers, all of them working at maximum capacity up to this point. This is one reason why Toyota's losses piled up so tremendously fast, turning the figures for 2009 into the worst ever – despite the fact that the Toyota fiscal year 2009 only covered two quarters of the financial crisis. There could have been no return to profitability without the drastic savings and the conscious shrinkage of the organization as a whole.

It is not just with an unexpectedly shrinking demand that Toyota finds it difficult to come to terms; if sales go up to unanticipated heights, the TPS also has trouble adapting due to its lack of spare resources.

Long waiting times for the Prius

Just six months after the second generation of the hybrid pioneer, the Prius, hit dealerships in Europe and the USA in 2003–4, waiting times for the Prius II rose to almost a year, especially in America, where the Toyota flagship became a sales hit. Demand had exceeded Toyota's expectations by no small numbers, and the organization simply could not ramp up production fast enough. By the end of 2004, Toyota had produced 120,000 Prius II, but demand in the US market alone hit 60,000. Another bottleneck might have occurred with the supply of the nickel-metal hydride batteries.

Environmental-friendly consumers in California, however, were willing to wait an entire year for the car – contradicting the perception that Americans are generally impatient customers who want to jump in the car and drive it straight off the dealer's lot. Hybrid fans with cash to spare even paid up to US$6000 extra for quicker delivery of the world's second mass-produced hybrid vehicle.

For a long time, Toyota was happy to accept this structural lack of flexibility, remaining as it did completely devoid of negative consequences – as long as Toyota was growing. It took the outbreak of a world economic crisis to show where the weaknesses lay.

Another difficulty is that Toyota can only draw on the full strength of the TPS at home; abroad, the *heijunka* principle of leveling off fluctuations by transferring resources and productions between factories is not practicable in every case.[13] In Japan, where the bulk of Toyota's core facilities are centered around Toyota City, a plant can remove or add staff without further ado, or send production of certain models on to another line or even to another factory, because most factories and suppliers are clustered around the city. In Europe, for example, this is only possible to a lesser extent, because factories are strewn across the continent rather than concentrated in any one place. Not only does this hinder the exchange of resources, it prevents production from being redirected, since most plants are more or less set up for the production of specific models. Sending workers to where they are required is difficult, too, with Toyota plants in six distinct countries: Great Britain, France, Turkey, the Czech Republic, Portugal, and Poland.

In order to better compensate for variations in demand and to use its Japanese sites to full capacity, Toyota had tried to link plants abroad more closely to production at home: the "global link production system," as it was termed, aimed to couple factories across the world with one Japanese sister site so,[14] faced with a sudden increase in demand, the Japanese factory could produce the extra units while the plant abroad kept to plan – i.e. the Japanese sites would have picked up any surplus demand. Yet this strategy would only have been effective where sales were peaking; when demand fell suddenly, as it did at the end of 2008, the global link would prove absolutely ineffectual. At present, Toyota is restructuring its production network, with the aim of introducing greater flexibility, especially in its home base Japan. The new key words are "stop and consolidate." The natural disaster of March 2011 has meant a radical turn-around on this point, with Akio Toyoda assuring Japan that there would be no cuts in home production – both in a gesture of national solidarity and in a desperate attempt to keep production from dropping any further than it already did in the immediate aftermath of the natural catastrophe. Further to this, there will be new investment in the Tohoku region in the north east of Honshu Island, the area worst hit by the disaster.

Factories in the homeland will specialize in new vehicles to be introduced, cars with new technology and innovative, low-cost production systems. Lower-tech products with less efficient production – mainly aimed at the export market in any case – will be produced abroad, making them less vulnerable to the rising yen. Other Japanese rivals have already made similar moves: Honda, for example, has opened up plants in Thailand that produce particularly for export. In terms of functionality, there can be substantial differences between the Toyota factories in Japan and those abroad, especially in terms of rigor in implementing the TPS. A prime example of sometimes dysfunctional TPS elements is the *andon* cord, which many European employees dislike pulling, especially if the production cycle is tight and output has fallen below target. Despite this psychological pressure, however, pulling the cord should always be the unquestionably correct response whenever a quality issue has been identified. But Toyota has discovered that this instrument of "built-in quality" can be turned into a disciplinary measure at European sites, to assist group leaders to keep tabs on their underlings. While employees in Japan are evaluated collectively based on team performance, in Europe the tendency is far more toward individual appraisals with no small degree of subjective opinion. This makes European workers far less willing to pull the *andon* cord, and whether they pull it or not has more to do with the wishes of their team leader than their own quality judgment.

The *andon* cord is a part of Toyota teamwork that is very much a part of Japanese culture. It functions without fail in its home environment because the individual does not risk loss of face by giving the *andon* signal. In Japan, a team leader does not automatically assume that a single employee is at fault when *andon* has been pulled, but first looks for problems in the process. In a Western production environment, however, mistakes are considered more personal in nature and at worst can lead, not just to loss of face, but to loss of employment. Meanwhile, in Japanese production culture any shame is shared by the whole collective – and only if and when actual defects are found on the vehicle, i.e. if the collective has not succeeded in averting problems during production. Toyota managers have admitted that *andon* does not always work well in the West, especially if the workers involved were not trained at Toyota, but come from other, Western manufacturers with different production methods. In most regions, Toyota is forced to use local workers, with Japanese Toyota staff only available for plant and regional management.

As soon as Toyota's ambitious *kaizen* philosophy clashes with the unavoidable adaptation to local culture, there is a real danger that the TPS, or at least crucially important elements of it, will no longer function correctly. Having said that, provided that the workers abroad are trained early – ideally at the beginning of their professional lives – and fully inducted into the Toyota production culture, there is no reason why tools like the *andon* cord would not work in, say, Europe or America and lead to the expected gain in quality and efficiency.

Then again, even at home, the TPS and its effect on staff are not entirely uncontroversial. According to Kamata Satoshi, one of Japan's leading investigative journalists, the intensity of work in Toyota factories on home territory was at times so high as to induce a rise in depressive illnesses and even suicides among employees. He quotes a report by the Toyota Labor Management Council (dated November 27, 2003) that describes the "high number of psychological illnesses as a worrying sign." This – and further reports like it – shows a dark side to the Toyota Way, a side that several authors who have written about Toyota's outstanding efficiency might well overlook in their depictions of Toyota corporate culture as "people-friendly."

There is no shortage of indicators that, in terms of human resource management, the Toyota Way is also a particularly smart concept that shies away from measuring itself against strong unions and, in part, forces its workers to adopt its philosophy. Certainly, employee representation at Toyota City is a million miles away from what is practiced in the Volkswagen home town of Wolfsburg. Toyota has enterprise unions, which in turn follow company strategic policy; and in contrast to the Western understanding of employee representation, they are ideologically neutral. In fact, those internal unions function more as the primary way for employees to contribute to improving processes: a full-time permanent Toyota employee, after all, has a life-long job guarantee whatever the case, with fluctuations in demand in Toyota City being covered more or less completely through temporary workers whose numbers have risen into the thousands during boom phases.

While the performance-related pay system has been modified during the last couple of years (which will not be further pursued at this point), what does become clear is that the interaction between *kaizen* and remuneration is a cornerstone of performance management, which stands in stark contrast to Western production culture. This strictly performance-dependent system, however, is not without pressures on staff, which is why it had to be amended. The restructuring, which has been called "the humanization of the TPS" by some authors, aimed to give more priority to the individual, and Toyota has adopted a firm-wide motto – "Developing people" – which, according to its self-image, is a top priority above everything else, even above product philosophy.

Whatever the case, in Toyota City the wellbeing, the welfare, and the prosperity of the employees very much depend on company gains in productivity; for many workers, this is a life-long relationship. Toyota's employees do not think twice about the crossover between the company and their personal and private lives, and in return, Toyota is a paternalistic provider for everyone who is permanently employed there. This gives the Toyota system a monolithic character and, even in Japan, it has been criticized for having a cold facade and a tendency toward shutting itself off. On the other hand, the strong identification of its workers with the car manufacturer is one of its most powerful resources.

3.5 A GLOBAL PLAYER WITH A DISTINCTLY JAPANESE IDENTITY

Two principles can be seen as exemplary for the global expansion of the Japanese giant, principles which Toyota has consistently employed to a greater or lesser extent at all of the more than 40 plants that the Company currently operates outside of Japan. They show the remarkably systematic methods that Toyota used to get to the top of the industry, one principle having to do with the conception and design of the factories, the other being a question of interpersonal organization between the Japanese managers and their workers at sites abroad.

In order to conquer foreign markets, Toyota used what is referred to as "transplants", factories which are conceived at headquarters in Toyota City and then transplanted abroad, the designs of which are exact enough for them to be put up at high speed and readied to the last detail abroad. Instead of reinventing the wheel with every new factory, leaving local management the freedom to tackle the process problems arising – as VW has been doing until recently, whether in Kaluga, Russia, or Chattanooga in the USA – the Japanese production network replicates itself continuously following an exact blueprint. While there may be the occasional deviation – such as in Great Britain and France, for example – in general Toyota has never digressed from this strategy. Transplant sites like those in the emerging markets allow Toyota to keep the number of management personnel from Japan required down, with processes in Toyota factories standardized almost down to the breathing of the worker. Indeed, at the Kolin factory in the Czech Republic, for example, where Toyota is cooperating with Peugeot PSA to produce the Peugeot 107, the Citroën C1, and the Toyota Aygo, observers report that processes have been designed to be almost completely error-proof; in contrast to, say, Volkswagen, there is almost no room for human error. Each movement has been preconceived, every step or stroke of the hand carefully planned in advance. Industry insiders do not shy away from strong terms like "idiot-proof" when describing Toyota processes, and predefined standards on all Toyota models reduce the number of variants and the amount of work required on the assembly line.

This level of planning and standardization makes the individual worker at assembly-line level essentially interchangeable. Furthermore, the work to be done is comprehensively visualized. Toyota leaves nothing to chance and, should a problem occur, it has institutionalized its arsenal of highly effective analytic systems (*andon,* the five whys) and applies them until the process concerned has been sustainably improved. While such processes may not always function as well as planned in foreign cultures, they are not without effect.

The structure of interpersonal relationships within Toyota staff is often transplanted into the location abroad along with the factory. Among the traditions which rule social life among Japanese staff can be, for example, the "boss wife": this tradition requires that the wife of the president of the factory abroad keeps a close eye on the family relationships of the Japanese employees at the plant. Should she hear of unusual circumstances that are leading to social conflict or family issues, the employee concerned (or, in some cases, his/her partner) must appear before her and be questioned about the situation. The rigor of "boss wives" is so legendary, and the fear of losing face within the organization so great, that most personal conflicts are nipped in the bud at an early stage.

In this way, Toyota's social organization supports Japanese members of staff who are cut off from home and have to cope with a foreign country and an unfamiliar language. Toyota always tries to minimize harmful effects and stays true to its traditions and heritage, wherever it sets up shop abroad. This close adherence to company principles is one of the driving forces behind the speed of the international expansion, but it also increases the difficulty (for the company) in adapting to widely heterogeneous and yet vitally important markets such as Europe, where the success both of Toyota products and of Toyota production methods have trailed behind expectations for years.

Then again, the transplant concept, which was used partly in the USA and in many emerging economies, for example, must be seen as a competitive advantage; without it, it is difficult to conceive how Toyota could have increased its worldwide sales by half a million vehicles per year. Volkswagen has set goals through 2018 that will also demand this speed of growth. Despite its success in 2010 and 2011, Wolfsburg has yet to discover a convincing answer to Toyota's astonishing ability to successfully replicate its production network almost anywhere on the planet and still keep costs down and processes stable and efficient.

In the meantime, Toyota has managed to find time to develop a very different project: the fully automatic factory. Automation of all sorts plays a considerable role in Japan, and the worldwide image of the Land of the Rising Sun as a country of *tamagotchis,* automatic toilets, and advanced, personalized robot technology is by no means a *cliché*; automation is beyond doubt a genuine strength of the Japanese economy and has a correspondingly important role in car manufacturing.

Opened in summer 2007, the third generation of production lines at Toyota factories is a shining example of efficiency and a textbook case of slim and lean production methods. The technological refinement with which Volkswagen invests its vehicles, at Toyota is dedicated to the design and innovation of the factory and its production processes: the new Takaoka line is matchless Toyota efficiency in its most tangible form. Instead of a transfer bar, robots work the stamping shop, the plastic molding shop, and the paint shop, cutting lead times, logistics, and assembly times by half. Through automation Toyota made Takaoka the fastest-producing factory in its stable. The new assembly line is reported to be just half as long as other production lines at Toyota, and to move 1.7 times faster than usual lines. A new painting process allows three coats to be applied at the same time, without having to wait for each coat to dry, and is something other manufacturers are following Toyota in adopting. This shortens painting times by 40 percent, cutting energy costs. Takaoka goes beyond visual inspection: for built-in quality Takaoka uses high-precision instruments to measure several parameters. The testing devices are located at various key stages on the assembly process and provide data in real time to factory managers and suppliers, minimizing the defect rates at all stations. Furthermore, the new Takaoka line was designed to give Toyota more flexibility than ever before: each line is supposed to produce eight different models, so the plant will produce 16 models on two lines compared with the four or five it used to produce on three lines. Toyota produced 220,000 cars per year per line in previous facilities; now at Takaoka, they are aiming for 250,000, achieving a higher variance (of models) with fewer staff.

If Toyota were able to implement the new Takaoka line on a worldwide scale, this would most likely open the door to the next evolutionary stage of the productivity-obsessed company. Yet currently, Toyota's crises have brought the continued roll-out of this kind of high-tech production to a halt, so this kind of competitive advantage has had far less impact than Toyota must have hoped. New lines are mostly being installed in emerging markets, for example in Brazil or China, where a high standard of automatization in plants does not pay off. For instance, Toyota's Chinese facilities, with their complicated joint venture-based corporate structure and comparatively cheap labor, are perhaps not the ideal place to deploy fully automatized assembly lines.

In terms of products, however, Toyota follows an entirely different philosophy and, with the exception of its environmental-friendly powertrain technology, is more of a fast follower than an innovator. Toyota's vehicle development is generally based on very small doses of technological innovation: unlike Volkswagen, for example, Toyota relies on an evolutionary, not revolutionary approach when it comes to new models, building cars that are easy and quick to assemble; anything else, after all, would be *muda*. The adaptations of the production process which the launch of a new model might require are far more important for the Japanese than for the Germans, for example, and can even influence the design of Toyota cars. New models are tried out in certain test regions at first and put to the proof for commercial success and customer satisfaction. This allows Toyota to learn and then roll out the product to new markets as soon as it has gathered the experience it needs for larger-scale production, avoiding high start-up costs and, in the long run, enabling remarkably good profit margins. While competitor Volkswagen often lets itself get bogged down in a fanatical attention to technical detail until the very last minute before a product launches, Toyota would consider that as pure waste and instead concentrates on stable, effective processes. In contrast to Volkswagen, Toyota offers variety in the form of many different models in the lineup, rather than variance or the opportunity to customize individual models. Then again, the sheer number of series running all the way down the scale at (very) close intervals leads to so many different models in different regions that Toyota has resorted to displaying the choice as a matrix.

This strategy has much to do with the fact that Toyota is always trying to keep the complexity of its products to a minimum, saving production costs. Volkswagen customers, however, can have cars such as the Golf delivered in more than three million different variants, creating huge costs for VW in terms both of assembly and of stocking (spare parts have to be stocked for at least ten years). This impacts the profitability of Wolfsburg models, while Toyota models are far less complex and offer far fewer optional extras to choose from, with corresponding effects on the Japanese company's profit margins. Furthermore, this simplicity has allowed Toyota to cut the development time for new models (from design decisions through to finished prototype) from 40 to 24 months. This makes drawing-board time for Toyota models among the shortest worldwide, and the Japanese giant had been aiming to decrease this even further to 18 months, possibly saving another 15–20 percent in costs and time. This would have meant that, including a follow-up period of one year for suppliers, new models would have been production-ready in just two-and-a-half years, setting the bar high for the European and American rivals. However, in the wake of the quality issues on the US market, Toyota publicly stated in July 2010 that it will allow an extra four weeks for model development in the future. Eighteen months might seem too aggressive even for the Japanese; however, in view of the Toyota ethos, the original target might well be restored some time in the future.

In contrast to most European competitors, for example, a change of vehicle generations at Toyota means that only about 30 percent of the new models actually have to be developed from scratch. Following its "carry-over" policy, in which 70 percent of new models must be taken from their predecessors, Toyota lags one or two vehicle generations behind other major manufacturers such as VW in terms of vehicle technology. This helps it avoid disturbances on the assembly line, since little has to be changed when a new model is to be produced. This strategy allows Toyota to continually reach the top of defect-free productivity rankings (see the sidebar on the 2010 J.D. Power plant awards in Section 3.4.2) even using generally older factories and tools that have long since been written off in the account books.

The more parts are standardized, the more stable the processes on the assembly line – and this is reason enough for Toyota, obsessed as it is with process, to let its competitors do the legwork in terms of innovative technology such as advanced driver assistance systems. Toyota tends to introduce innovations like Electronic Stability Control after its competitors, with the Prius being once again the exception to this rule. This may well be another reason why Toyota still faces difficulties in European countries, where exactly this kind of technological advance is needed to distinguish and strengthen a brand in these otherwise saturated markets.

Toyota offers a large range of models, but Toyota's car design has often been perceived as "conservative," particularly by Europeans. The kind of branding done by Volkswagen, in contrast, which allows single models to develop their own product identity and position themselves on the market with an individual profile, has long been irrelevant for Toyota (with the exception of Lexus: see Section 3.6 – Toyota's luxury strategy: *Lexus vs. Daimler and BMW*). Even in the sports segment, for a long time the company had been reluctant to offer *avant garde* products; models like the smooth and youthful Celica were what Toyota produced. The Japanese followed their own tried-and-tested logic, meaning that sales would automatically pick up with rises in the quality and durability of the product. Even Toyoda's predecessor Watanabe was convinced that the carmaker should concentrate simply on extremely high quality, and that sales would follow of their own accord.

Toyota struggles with the concept of brand identity or individual model personality as far as characteristics other than pure product quality and environmental-friendly powertrain technology are concerned. The effort Toyota made in rebranding its best-selling Corolla in Europe is an obvious example. The model, renamed the Auris in 2007 and supposed to position itself against the VW Golf, flopped dramatically in many European markets. Surprisingly enough, even the million-euro marketing campaign unleashed by Toyota (27 photos, 200,000 poster spots booked) entirely failed to entice customers. The Auris – just like its predecessor the Corolla – offered a somewhat indistinguishable design and faded into the background of the highly competitive European compact segment, instead of catching customers' eyes. In Germany, for example, the Auris was one of the reasons why Toyota lost market share in 2007, selling only in the 20,000s in the year of its release and slipping to 16,000 by calendar year 2009[15] in Europe's largest car market.

Toyota's design motto, "vibrant clarity", is intended to display a rational, environmental-friendly product philosophy, and to date Toyota's ED2 – or European Design and Development Center, near Nice in southern France – has not been able to add the kind of flair to this philosophy that makes cars attractive to style-conscious Europeans. Furthermore, since the introduction of the Prius, Toyota has increasingly defined itself as a pioneer in the hybrid segment, and it is in the Japanese and US markets that it has self-confidently taken on this role – but not in Europe. Nevertheless, in the eyes of the Toyota giant, environmentally conscious manufacturing and the contribution this makes to protecting planet Earth is the best marketing image a car manufacturer can hope for.

The industry's largest manufacturer is only just starting to adapt its product philosophy in order to pep up the emotional appeal of its models. The new sports car study FT-86, for example, indicates that the Japanese company is slowly starting to get to grips with the importance of emotion in car-buying decisions, with this vehicle concept showing a far edgier and more sporty exterior than its predecessors. This change of course, which Toyota will be pushing more than anything in Europe, was worth being outlined publicly when Toyoda presented the 2010 annual results: "We want to provide customers with vehicles that emphasize driving seasoning and the joy of automobile that customers find exciting, such as the compact FR sports car, FT-86 concept, unveiled at the Tokyo Motor Show last year." The extraordinary choice of words – "driving seasoning" – may be proof of the fact Toyota is still very much at an early stage with this new strategy and finds it difficult to speak about cars in such an emotional way. The new sports car is reported to be due for release in 2012.

CEO Akio Toyoda, who refers to himself as passionate about cars and loves driving in rallies, is doing his utmost to change this historic lack of communication. One of his tasks is to tool up Toyota vehicles with intangible values such as fun and a distinctive design language. Help is expected from partners such as Tesla. This carmaker offers sporty, environmental-friendly vehicles that can go from 0 to 60 in record time – all without putting a gram of CO_2 into the atmosphere. This is just the mix that Toyota is looking for, and the small electric segment is just the place to try it out, with electric motors having a clear advantage over combustion engines in rate of acceleration. This is why, at the press conference announcing Toyota's taking a stake in Tesla, Toyoda did not tire of stressing just how much fun he had had with Tesla boss Elon Musk driving the Tesla Roadster, which has become famous well beyond the borders of the USA. It remains to be seen, however, whether the Japanese giant can get rid of its genetic tendency to prioritize the inoffensiveness of a model over its emotional appeal. (See Figure 4.1 in Chapter 4.)

3.6 TOYOTA'S LUXURY STRATEGY: LEXUS VS. DAIMLER AND BMW

Despite constant claims at Toyota City that Toyota does not measure itself against its competitors and does not want to "beat" them (Toyota has, to this point, always remained true to itself and refrained from crowing about its success), the German premium manufacturers Daimler and BMW were role models for the Japanese over decades, who looked up to their dominance in the luxury segment in USA, which is approximately only half the size per total volume of Europe. Toyota respected the Europeans' technological advances and their premium quality standards and, eventually, decided to try to catch them – even though for years, the idea of a Toyota luxury model had seemed somewhat far-fetched.

The driving force behind the decision to enter the luxury market was Toyota's most important market, the USA. More so than in Japan, Toyota compared itself to rivals Honda and Nissan when looking at the USA, where the other two Japanese car makers always seemed to be one step ahead in terms of implementing successful strategies in this as yet unfamiliar market. Honda, for example, far smaller than Toyota, took its production to the USA long before Toyota and had already made a name for itself with its fuel-efficient engines. The Japanese giant kept a close eye on these developments and not only followed suit with fuel-saving technology but also wanted to rid itself of the image it had at the time as "the mass market car that doesn't break down." Honda had gone ahead with the Acura,[16] and Nissan followed at the end of the

1980s with the luxury brand Infiniti; both took on the established European premium manufacturers for market share in the US premium segment which, compared to the European market, is small and therefore hotly competitive.

The new wave of Japanese carmakers with luxury designs on the US market at the end of the 1980s did not come by chance. At the time, vehicles in larger classes could be imported into the USA despite the voluntary trade constraints the Japanese had placed on themselves; even where the import of the finished car was not permitted, the parts could still be shipped to the USA and then assembled there. At the time, however, bilateral treaties still placed strict limits on Japanese cars in general, especially in the smaller vehicle classes. So Toyota circumvented this by importing the luxury Cressida/Corona Mark II, which was one class below models made by competitors Daimler and BMW and became very popular among the baby boomers. Toyota recognized that they would need a strategy to retain this group as their incomes rose with their ages, and for this reason developed its own luxury brand, aiming to combine the functional amenities of the vehicles from southern Germany with the quality and reliability synonymous with Toyota – all of this was to be offered at a modest entry-level price.

The prototype for this new premium brand, for which the brand name Lexus was quickly coined, was the Cressida series. The design, however, was more oriented toward European manufacturers and, as ever, the Japanese started small and took modest steps, collecting experience as they went along before moving to attack BMW and Daimler. The first Lexus LS was sold in the USA in 1989. Toyota was aiming to skim off the then high profits in the premium segment, with top-of-the-range models at the time yielding a far greater share of the overall profits carmakers could reap than the compact segment – and in any case the company had already done well in that segment with the Corolla. In addition to the profit motive, however, the Lexus was important for Toyota because, for the first time, the manufacturer was going beyond pure functionality and offering car buyers a vehicle that served as a wish-fulfillment item, promising nothing less, in fact, than the embodiment of the American Dream. By the same token, it was also the perfect antidote to Toyota's nightmare of staying locked in the good-quality, low-price category and showed that the carmaker was capable of moving into premium territory.

What Toyota started with the Lexus can be seen in hindsight as the birth of active brand management at Toyota City. From its inception, Lexus's success story is a textbook case of clever marketing strategy, with Toyota starting a separate sales system in order to keep the brand completely uncontaminated by the mass-market segment. After all, the new luxury flavor required a completely different image than the no-frills cars Toyota had been successfully selling so far. Still, the Lexus was a thoroughly Toyota product in terms of its quality standards, which is how Toyota distinguished the brand from the competition. Rather than trying to outdo other premium cars in horsepower or dynamic driving qualities, Toyota paid attention to USPs that had less to do with aggression, sport, and speed and more to do with quality finishing. In any case, powerful motors and high velocity seemed less important for both the US and Japanese markets (the top two Toyota sales regions) due to strict speed limits on most highways. Instead, the Lexus brand is based on top-of-the-range production techniques (e.g. they are sandblasted six times) and, simply put, perfectionism. In the Miyata Lexus factory in Fukuoka, Japan, for example, there are workers employed to stroke the bodywork with their hands, looking for any irregular spots and checking for gaps between the sheet metal components. Wearing spotless white gloves, these employees are trained to feel discrepancies down to a tenth of a millimeter; to outsiders, it seems that they are lovingly caressing their cars with the same devotion as a Japanese Kobe beef farmer massaging his cattle.

By the turn of the millennium, Lexus was the undisputed king of the American premium segment, trailed by its former role models Daimler and BMW.

Market leader in the USA

In 2005, Lexus hit sales figures of about 300,000 and made it into the provisional record books for the luxury segment. It sold about half of this number as passenger cars and the other half as SUVs.

Through the first ten months of 2010, however, it seemed that for the first time in a decade Mercedes-Benz was close to toppling Lexus. The German premium manufacturer had sold more than 184,000 vehicles in the USA, about 1000 more than Lexus, according to Autodata Corp. BMW was about 7000 vehicles behind. The fight came against a backdrop of improving auto sales for the luxury market in the USA. Luxury auto sales were up more than 13.2 percent that year compared with a 10.6 percent gain for the entire industry, according to Autodata. Lexus had to battle headwinds from the recalls of millions of Toyota vehicles for sticking gas pedals, faulty floor mats, and electronic and software issues.[17] In the end, however, the Japanese brand finished the year as the number one luxury car brand in the USA.

Before even launching the Lexus brand in 1989, Toyota had started to look for dealers early, conducting a careful search and literally handpicking their sales partners. This was another major advantage over competitors, who were often entangled in a web of sales structures and contractual obligations after decades at the top. Over twenty years later, much of Lexus's success can be traced back to contented dealers selling this high-quality product, and when the competition started seriously analyzing the Lexus strategy just a few years back, they found that the brand had the highest new-vehicle throughput per dealership of all US car franchises, with the comprehensive service offered by dealers in terms of after-sale service and repairs (very crucial to the US market in contrast to elsewhere) forming the cornerstone of the consistently high customer satisfaction. Lexus even recovered somewhat faster than the Toyota brand from the recall issues in the first months of 2010.

Toyota gave itself time to learn the lessons that came out of the Lexus experience in the USA before setting its sights on Europe. It always takes high-level strategy decisions in a conscientious and careful manner and, after showing the Lexus at the 1990 Geneva Motor Show and introducing it on a trial basis into the European market, Toyota waited for quite some time before really pushing the brand there. It was not until 2005 that Lexus was comprehensively rolled out in Europe, and despite the high-level planning efforts made, the brand has not been able to establish itself on the home turf of the world-famous German luxury manufacturers.

Despite the fact that almost all Lexus models are produced in Japan and exported, more than anything else the brand caters to American taste – a possible reason for Lexus trailing behind in Europe with just 26,000 cars sold in fiscal 2010. In the Japanese luxury market, Toyota's Lexus strategy is slowly starting to succeed, even though it had only been introduced in 2005. Before this, Lexus models had been sold under the Toyota brand name. Like the US strategy, Toyota Japan also opened up new sales channels (known as Lexus Customer Lounges). In the same year, Lexus entered the Chinese market, where sales totaled 100,000 in 2009 – China is a promising market in terms of the long-term growth prospects. With the Global Vision, Toyoda proclaimed that Lexus should establish a broader basis in emerging countries such as China – Toyota obviously has recognized the sales potential for luxury brands in China.

While Lexus uses synergies with Toyota's high volumes wherever it can, as a general rule Toyota introduces new technology on Lexus cars first, rather as Volkswagen does with Audi. So in order to give Lexus a green profile and to give it another universal selling proposition to distinguish it from Audi, BMW, and Daimler, Toyota did not hesitate to equip its luxury models with hybrid powertrains. The RX 400 SUV was the first vehicle in the luxury segment to use hybrid technology, with fuel consumption of just below 30 mpg) – i.e. not much more than a sedan with a standard four-cylinder engine. Nevertheless, it is hard to see the Japanese premium brand as a danger to the European manufacturers in their homeland, despite increases in fuel prices and paradigm shifts in climate change policy; in Europe, environmental-friendly powertrain systems have consistently failed to go mainstream. European customers, nevertheless, seem to have a strong taste for luxury cars that are especially designed for fast highway driving while maintaining high comfort – a specialty of Audi, Daimler, and BMW.

3.7 MOTORIZING EMERGING ECONOMIES: TOYOTA'S "CAR FOR THE WORLD"

What may be considered a weakness for Toyota in European markets, however, is a crucial source of strength in the emerging economies, with Toyota having shown a much better understanding of what wins over customers in Asian countries and other rapidly industrializing regions than its competitors; and the company is eager not to lose touch here. The Etios, for example, unveiled at the Auto Expo 2010 in New Delhi, is a compact car with a lot of space and an economical engine specially developed for the Indian market. With its rounded form and friendly face, this car distinctively caters to Asian taste and Asian customers, who – in contrast to Europeans – in general do not prefer edgy, sporty, or aggressive designs, all the more so in the compact segment.

For a variety of reasons, Toyota has long held the motorization of the emerging economies as one of its central goals. Like other manufacturers looking to move into these markets, Toyota has had to find ways around punishingly high import tariffs, and as a result started to build factories in Brazil and the ASEAN states as early as the 1950s and 1960s. Back then, there was another reason for this expansion into today's top-growth markets: at the time, Toyota did not see the quality of its vehicles as high enough to break into European or American markets. In Asia or Brazil, however, the only competition Toyota faced was from other Japanese brands such as Honda, Suzuki, Nissan, and Mitsubishi – and just like its home-country competitors, Toyota was able to develop the kind of cheap, robust vehicles these countries required.

Volkswagen has always had trouble offering low-tech models that can be sold at an affordable price in markets such as South-east Asia, the Middle East, South America, and Africa, with Brazil being the only real exception to this rule to date (see Section 2.6); Toyota, however, has for years now been supplying these markets with cars that lack expensive technical add-ons or interior fittings but can be produced cheaply and profitably. This strategy has given Toyota a respectable market share in these markets: in Asia, (excluding Japan and China) for example, Toyota sold almost 1 million vehicles in fiscal 2010. This shows the strength of the manufacturer in these markets in contrast to Europe, where it sold just 858,000 cars in the same period – nearly 150,000 fewer. In the Global Vision 2010, the role of the emerging economies became an even more important part of Toyota's former strategy of reaching a 15 percent share of the global market, and this was in no small part due to the poor prospects of getting out of low single-digit figures in Europe. Toyota's target for 2015 is now to achieve approximately equal weightings in unit sales between industrialized nations and emerging markets. In emerging markets the company wants to partake in the growth in demand by concentrating on locally-produced core models, including the innovative International Multipurpose Vehicle (IMV) models and newly developed sub-compact models.

The IMV series started development in 2002. Toyota took up the challenge to develop a "world car," internally termed "innovative international multi-purpose vehicle." The Corolla had already been successful in several emerging markets, but Toyota decided to start afresh with a new platform from which five models emerged: three pickups, a multi-purpose vehicle (MPV), and an SUV. The core target areas for the series were Asia, Eastern Europe, Oceania, South America, and the Middle East. This was where Toyota was most concerned about the rise of its Korean competitors, who were already selling particularly good value vehicles in these markets. Toyota models, meanwhile, had crept into higher categories beyond the average purchasing power of many consumers there. Honda, too, was hot on Toyota's heels, and the IMV was intended to fill the gap and push forward localization outside of Japan in order to keep costs down to levels comparable with the Korean competition. For the first time, Toyota shifted some of the development of this new series to emerging countries, the bulk of its components being sourced and produced there, too.

Toyota did not spare the expansive language when it announced the project, referring to it as an "industrial project that spans the globe"; and the logistics certainly produced challenges on a global, if not astronomical scale. To meet them, Toyota drew on capacities in four factories spread across three continents as well as much of its supply network. Coordinated from Toyota City, the series was partly developed and manufactured in Thailand, with parts coming in from Indonesia, the Philippines, Argentina, India, and Africa – and all of this sticking to Toyota lead times and just-in-time processes. Yet despite this almost insurmountable task, Toyota's world car was a success: in the fiscal year 2008–9 – a period marred by the outbreak of the financial crisis – the IMV series sold 700,000 units worldwide, a figure 21.6 percent above comparable sales for the previous year. In Asia, Toyota booked strong IMV sales, and the net turnover in the Asian region outside of Japan and China went up by 40.2 percent, climbing to JP¥31.209 billion (US$335 million, €233 million) and giving an operating profit of JP¥2.564 billion (US$27 million, €19 million). As early as mid-2010, exports of Toyota's IMV series manufactured in Thailand alone were reported to have touched the 1 million mark. In addition to this, the factories in Indonesia, South Africa, and Argentina producing the IMV were able to continuously increase their production.

Despite these successful figures, the series – intended to offer economic motoring for emerging economies – ended up developing in an unforeseen direction in many markets. In Thailand, for example, the Fortuner IMV has climbed out of its price category and target group into a higher segment, and is considered a vehicle for high earners. So Toyota turned the innate momentum of this series and its tendency to follow its customers' growing purchasing power up its segments into a strategy. In 2007, the former CEO Watanabe went on record as saying that "Toyota can never be a cheap brand; it is a quality product with a fair price, which in emerging markets may be a premium price. We may not necessarily be somebody's first car; we definitely want to be the second car that the family buys."[18]

Despite this, it did not take long for competitors such as Hyundai to show the world's largest carmaker just how much this "first car," produced at competitive costs, could drive large gains in market share; and Toyota, which is still having trouble producing cars that sell below the CN¥70,000 mark (approx. US$10,000, €7000) in China, for example, set itself to recapture the bottom end of the market. India is the first step in this strategy: in 2010, the second Toyota plant in the world's largest democracy began production, and the new model produced there, the Etios, shows that Toyota understands the importance of localizing design and tailoring it to developing markets. The Etios compact is the first Toyota that is specifically conceived to be competitive in the low-cost compact segment, which is showing high growth rates in India. Selling for less than US$10,000, this is probably Toyota's least expensive vehicle. In India, the highest growth rates are believed to be found in the segment below or close to US$8000 (€5500), and the carmaker wants to tap into this and sell 70,000 models before 2011 is out. In 2012, Toyota wants to roll out the Etios in Nepal, Bangladesh, Sri Lanka, and other countries that drive on the left.

In India, the fight for market share in the low-cost segment has been heating up with a recent stream of new entries in the region: among others, the VW Polo, the Chevrolet Beat, the Ford Figo, the Nissan Micra, and a new compact from Honda are all trying to win over Indian customers. In an attempt to hold its own not just against these, but against strong, established rivals such as Maruti-Suzuki, Hyundai, and Tata, Toyota has gone down new and innovative roads. The concept work on the Etios takes account of the availability of materials in India, keeping import costs low and minimizing the distorting effects of exchange rate fluctuations: almost all of the body parts required for the car are sourced in India, only the engine itself and gearboxes coming in from Japan. Toyota has recruited around 30 new suppliers for the new compact and is already working on a diesel version to cater to the Indian preference for this fuel.

The Etios has been in the market since late 2010, and is intended to boost Toyota's low Indian market share of just 2 percent, showing that it is not yet a well-known brand on the subcontinent. The new compact also makes clear that Toyota is by no means content to leave the low-cost segment in India to its rivals; it is taking the fight to them. Toyota is calling this new strategic compact car "an unparalleled experiment in developing low-cost vehicles," and if it proves successful, this will light the way for the carmaker in China, South America, and other emerging markets. In September 2010, work began on a new Toyota plant in Brazil that should also be producing a locally developed, low-cost car by 2012. In India, meanwhile, Toyota aims to reach a 10 percent market share by 2015 and is planning to launch a slew of new models.

3.8 MANAGING RELATIONSHIPS THE JAPANESE WAY

Toyota has grown into a global giant that is present in every major world market; and compared to its German challenger Volkswagen, Toyota plays a far greater role in a whole series of small emerging markets due to its wider range of low-cost models – which are popular even in the most remote regions of the world. In 2009, Venezuelan president Hugo Chavez, for example, threatened to throw Toyota out of the country if it did not start work on a four-wheel-drive off-roader suitable for public transport in the poor, isolated mountainous Andean regions. This incident is not just proof of Chavez's somewhat bizarre understanding of the global economy, but of the importance of Toyota's presence in the emerging economies of the developing world.

None of this would have been possible without Toyota's special relationship with its suppliers, which allows it to roll out its exceptionally efficient production system over the entire value creation chain all over the world. Toyota achieves this by offering its suppliers a win–win situation, the structure of which we will now examine more closely.

In the past, when Toyota was still a small manufacturer, it had to take a lot of care to give its small Japanese suppliers the technical and financial support they needed to grow with the carmaker – no easy task considering the scarcity of materials and manpower in postwar Japan. The manufacturer responded to the challenge by cooperating closely with its suppliers in order to better defend common areas of interest. This was the beginning of the tightly knit network of companies, known as *kairetsu,* which still determines Toyota's supply and production in Japan to this day. In contrast to the Volkswagen Group, which often changes suppliers, Toyota establishes and maintains long-term, stable partnerships.

Toyota's wide supplier network means that it has to assemble very little in-house, and this is, in turn, one of the major reasons why Toyota can keep costs low. The structure and functionality of this network is still a field being researched, but the basic set-up remained unchanged for decades, comprising the manufacturer itself, its primary suppliers for car bodies, engines, and transmissions (also known as tier 1), and a second rank of suppliers for single parts (tier 2).

What is remarkable in contrast to Western manufacturers is the duration and constancy of the business relationships within this intricate network structure, composed as it is of highly specialized companies and based on a high division of labor. There is hardly another carmaker with such a detailed range of different suppliers, and no Western manufacturer outsources such a high proportion of its production as Toyota – with the highly specialized German sports car manufacturer Porsche perhaps being the only exception.

This makes the actual Toyota Motor Corporation itself far smaller than Volkswagen, which has over 360,000 employees: TMC only employs 71,116 people directly (as of 2010).[19] Even the entire Toyota group as a whole has about 320,000 employees on its books (Figure 3.1).

The Group consists of hundreds of companies; as of 2005, for example, it counted 520 corporate entities, with another 56 affiliated with TMC through stakes held by the core company, called "equity method firms", and a considerable portion of Toyota's profits came from what is titled in the balance sheet as "equity in earnings of affiliated companies." During the boom from 2004 to 2008, this item produced tremendous results, adding to the strong flow of profits into Toyota coffers: in 2008 alone, these affiliated companies added JP¥270 billion (US$2.9 billion, €2 billion) to net revenues.

Compared to Volkswagen, Toyota is at the center of a complex network of mutual equity holdings and crisscrossed stakes between suppliers that is stabilized by distinctly Asian methods of balance and control. Of the 50 to 60 companies on which Toyota can exercise decisive influence, the largest

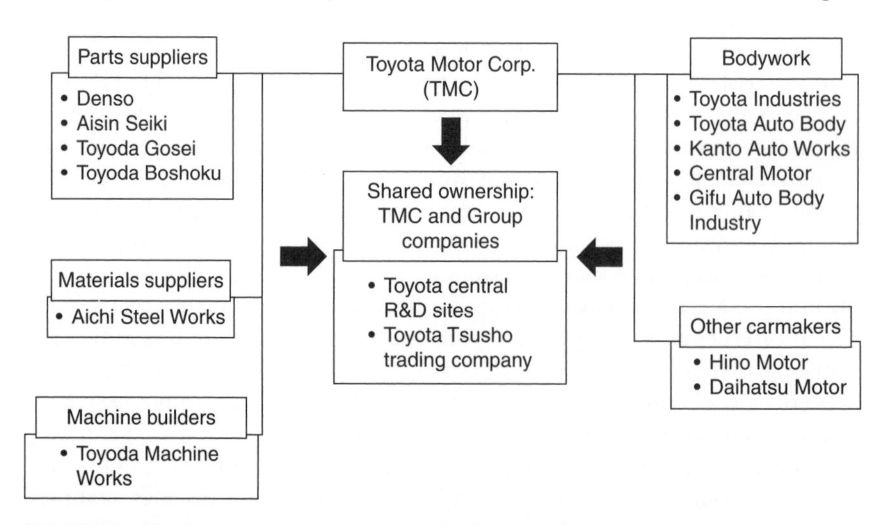

FIGURE 3.1 **The Toyota concern: a network of companies.**

Source: Ulrich Jürgens, "Warum Toyota so lange so stark ist", Wissenschaftszentrum für Sozialforschung, Berlin, 2007 (transl. "Why Toyota has stayed so strong for so long").

number are direct suppliers and companies doing commissioned work for Toyota. They can be divided into three groups as follows.

The first group is composed of brand-name manufacturers Hino and Daihatsu, both of which are listed companies and are financially consolidated by Toyota. Hino produces trucks, Daihatsu makes compact and subcompact cars. Both companies round off the Toyota product portfolio. They are the only two major acquisitions made by TMC in its entire history.

The second group is far larger, encompassing the around fifty firms which do commissioned work for TMC. These companies do not just take on vehicle assembly in its purest form (chassis building, body making), but cover the whole process chain from pressing through to painting and finishing. One example of this part of the network is Kanto Auto Works, an assembly specialist which has plants of a similar size to Toyota's and also carries out research and development for the parent company. Kanto is a listed company and thus has its own access to the capital markets. Additionally, its plant in Iwate at the north of Honshū, Japan's principal island, is one of the most modern car production plants in the entire Toyota stable. In July 2011, following the difficult period caused by the March earthquake, Toyota announced that it would be coming to the aid of Kanto Auto Works by converting it into a full subsidiary, along with another group company, Toyota Auto Body, making evident just how important these affiliated companies are for TMC.

The third group, referred to as *itaku,* do not produce own-brand vehicles, as Denso or Toyota Industries do;[20] most of them have a structure which mirrors that of TMC itself, consisting of a slim-line core company surrounded by a wide circle of subsidiaries and affiliates. Denso, the largest of Toyota's many suppliers, for example, has a core staff of approximately 39,000; including affiliates, however, it employs about 120,000 people.

In contrast to VW, around 40 to 50 percent of Toyota's vehicles are produced by outside companies. The whole Japanese production base consists of the five factories run by TMC itself and nine affiliated companies with around fifteen additional plants whose capacities are either fully or partially taken up by commissioned production for Toyota. Of these companies, 14 are bound to TMC financially and legally in some way or another; the parent company only owns five plants with finishing facilities.

The structure of the Group has a central function in the TPS inasmuch as it provides Toyota with a certain level of flexibility in its Japanese manufacturing base. In 2005, for example, this assembly network produced 53 different models, with several series being produced simultaneously in different factories. The exact combination of plants and production seems almost arbitrary, with combinations of TMC plants and other network factories for either purely TMC or purely commissioned production. In 2005, 40 models were produced in a single factory alone, 15 in TMC plants, and 25 in plants run by affiliates. The number of models produced per factory varied too, with two plants only producing two models and four producing four vehicles; Toyota Auto Body, which has two facilities, can produce as many as thirteen models. This keeps the capacity usage in the *kairetsu* network high and assures high productivity for the suppliers to balance out the often punishing demands made by the parent in terms of price and cost reduction. This balancing of capacity within the group is a key aspect of Toyota's *heijunka* principle and therefore of its competitiveness.

This network of Toyota-owned suppliers and affiliates has another crucial function: it allows for staff transfers between group members, and manufacturers that, like Toyota, offer employees a life-long job guarantee need some leeway to balance employment structure. In this way, affiliates function as a safety net for TMC core staff, and Toyota strengthens ties by sending managers out through its network, with roughly 120 Toyota managers per year transferring to its suppliers and affiliates. Many of the positions they hold in the companies in question are permanent positions reserved for Toyota people, and 11 percent of the managers of the most important suppliers are former Toyota executives. This is another indicator of the high degree of integration between TMC and other group members, a pattern which assures the parent company a high degree of power and control. Furthermore, the members of the Toyota network are also highly experienced practitioners of the Toyota Way, and the Toyota managers sent out to them make sure that the ethically grounded Toyota culture and its characteristic fanaticism with regard to quality is rigorously interpreted in all parts of the empire. Over the years, Toyota has implemented the TPS throughout its value creation chain,[21] from the suppliers right through to the first workstation at TMC. This assures consistent, highly competitive structures all the way down the chain.

Toyota even takes care of purchasing materials for its most important suppliers, sinking fluctuations in their risk insurance costs into the price of raw materials. In return for this kind of service, the suppliers in this network not only fully adapt their process management, but allow Toyota access to their accounts, their price calculations, and, of course, their production methods. Proceeds from efficiency savings that Toyota suggests must also be shared with the manufacturer. Companies that fully integrate into the system and play by the rules, however, are almost guaranteed a follow-up order in the next production cycle.[22]

Within these relationships, TMC's task is to set targets for productivity and to offer support in terms of *kaizen* and continuous improvement of processes and products. The supplier has the job of hitting targets, continuously raising productivity, and making every effort to support Toyota's processes in a consistent and stable way. The thick jungle of mutual stakes and management swaps between Toyota and its suppliers is often depicted as a concentric organization (see Figure 3.1), in great contrast to the vertical *kairetsu* preferred by other Japanese manufacturers in which the Original Equipment Manufacturer is at the top of a pyramid.

In the Toyota model, meanwhile, both sides can count on long-term relationships and therefore do not have to spend time making complicated arrangements in case a new product is to be launched or a factory set up. What is frequently described as something along the lines of "mutual agreement in launching new products or production systems" in the automotive industry – sounding simple enough on paper – is in fact one of the biggest challenges in actual production management, where thousands and thousands of parts need to arrive at the right time in the right configuration and at exactly the right place. This can be a big risk and potentially incur high costs if the two parties do not know each other and still have to establish communication links. The more often products are renewed and the faster the company is growing, the more important the kinds of tried-and-tested, long-term supply relationships become. For Toyota, they are one of the pillars of its enormous competitiveness (Figure 3.2).

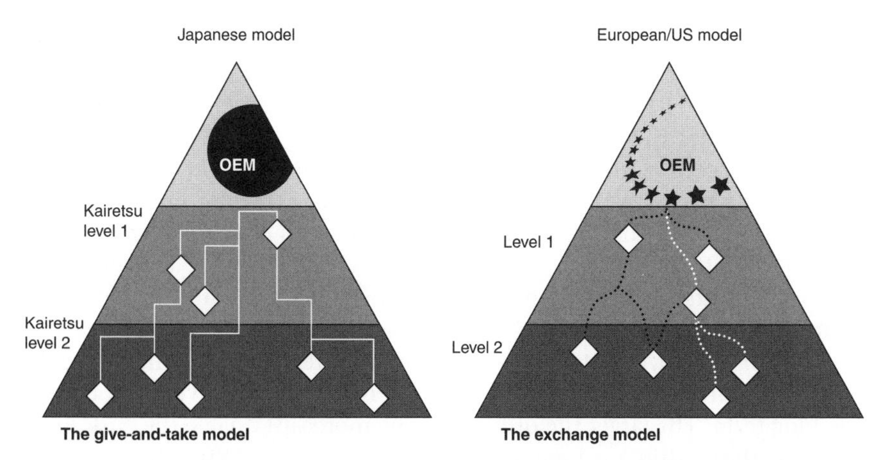

- High degree of financial interlocking between companies
- Cost advantages due to reduced process costs
- Comparative lack of motivation for suppliers to innovate compensated by the OEM itself
- The OEM pushes for continual improvements in production, leading to a higher degree of stability and quality.

→ Fixed network structures

- Tendering processes with a high number of competitors leads to variety and innovation
- Advantages in innovation due to high degree of R&D taking place at suppliers
- Reduced costs due to global sourcing
- Development service providers have an important part to play
- Contracts are often awarded to 2 or 3 suppliers.

→ Fluid network structures

FIGURE 3.2 **Supply management: Japanese versus European/US model.**
Source: PA Consulting Group.

Nissan's kairetsu

While Toyota's long-term supply agreements have been a consistent source of strength, the Land of the Rising Sun also offers spectacular examples of how these relationships can become dysfunctional. In the 1990s, Nissan went through a financial crisis and, in 1999, Renault took a stake in the company. Its first initiative was to examine the 1394-strong *kairetsu* of suppliers, who were partially blamed for the financial problems. When Renault compared the prices Nissan was paying to suppliers, the French car maker found a discrepancy of 20 to 25 percent. The comfortable and friendly relationship between Nissan and its *kairetsu* suppliers had led to clogged processes and uncompetitively high costs. The new Nissan CEO, Carlos Ghosn, took hold of the situation and, to the shock of the Japanese suppliers – and politicians – more than halved the number of companies in the network, amalgamating Nissan's supply with Renault in a common global procuring function for the two carmakers.

Even though in the worst case the Japanese *kairetsu* model can lead to a slackening of competitiveness, Toyota on the other hand has shown that it is not only able to hold its *kairetsu* together but also to exercise strict cost control.

Toyota applies the *kairetsu* system only at home, but Western suppliers that have succeeded in securing Toyota contracts often report that they, too, receive visits from Toyota managers who encourage them to make improvements in line with the TPS. Many of the companies concerned were astonished at the efficiency improvements resulting from Toyota's advice. Toyota certainly profits whenever it helps its suppliers, since it plans on long-term relationships. It is a classic win–win situation.

Furthermore, Toyota's tendency for long-term relationships allows suppliers to invest in capital-intensive equipment, machines, and even factories for the long term. The larger the orders and the more sustainable the customer–supplier relationship, the higher are mutual cost savings. When Toyota opens the door to new sales markets, suppliers frequently follow the company into new regions: wherever in the world a new TMC plant is built, there is generally a train of suppliers which set up shop nearby. Even in comparatively closed markets such as China, Toyota takes a number of suppliers with it, and this form of cooperation allows the carmaker to keep to its production philosophy and save costs in many countries in which it is producing.

In the West, however, the structure of an OEM's supplier management usually looks different. In the USA and Europe, carmakers see a new vehicle, a new factory, or a new component as a perfect opportunity to test the performance of different suppliers and to eventually choose the best performing provider. There is very little account taken of any investments the supplier may already have made, even where a supply plant has been set up in a new territory under an agreement that the carmaker will purchase there. Western carmakers such as Volkswagen take every opportunity to see just who can deliver the cheapest, and have no qualms about swapping suppliers, even for relatively marginal cost savings, since the potential costs or disadvantages of changing suppliers are generally considered to be minimal.

Toyota's challenger Volkswagen uses its calls for tenders to stage a harsh competition, with not just cost differences, but also innovations deciding the race between suppliers. The aim is to offer better, lighter, or more versatile products for the same price as before, and the supplier who makes the best offer tends to win the contract. This has the advantage that, to put it bluntly, the manufacturer in question can essentially lean back and wait for a tasty selection of new ideas to be dished up at a bargain price. It is the suppliers who carry the risk of investing in technological innovation that may never cover its costs, and there can be no question of a win–win bargain.

Set against the Asian model as practiced by Toyota, this leads to a far larger portion of innovations being developed by suppliers: in Europe, for example, the race to catch up in the field of hybrid powertrain technology is partly being shifted to the providers of drivetrain components. While Toyota developed hybrid technology on its own – and has therefore kept this technology very much to itself – in the West, the share of innovations coming from providers is rising. Bosch, for example, a huge German-based supplier and multinational corporation, provides OEMs with high-performance power electronics for hybrid vehicles. German-based ZF supplies hybrid transmissions – to Volkswagen, among others. Without suppliers' technological and financial strength, it would be almost impossible for Volkswagen to catch up with Toyota in hybrid technology.

In the wake of the March 2011 earthquake and tsunami in Japan, there has been no shortage of observers in the West who see *kairetsu* as one of the factors behind Toyota's production difficulties: the argument goes that Toyota's long-term reliance on its tightly knit network of Japanese suppliers meant that it could not switch to parts supply from other suppliers worldwide. Yet this criticism is facile, since a natural catastrophe of this magnitude would have caused havoc in supply chains operating on any model. The fact that the vast majority of Toyota's suppliers are located in (and, in the case of international operations, often run from) Japan is more to do with the way Japanese industry as a whole is structured than with *kairetsu* as a system of organizing a supply chain. Both the Japanese and the Western strategies have their advantages: Toyota's long-term approach allows it to profit from any cost-saving measures developed in cooperation with suppliers, guarantees smoother interfaces, and optimizes processes in a stable supply chain.

Volkswagen, meanwhile, changes providers more frequently, often achieving significant cost savings or technological gains by doing so and forming a strongly competitive landscape in the sector, the fruits of which it is then able to pick. This makes the number of innovations going into each new vehicle far greater. While Toyota's cost-cutting approach is particularly advantageous for companies enjoying stable growth, the innovation-led approach in the West can have advantages for manufacturers when components have to be changed frequently or when market risk posed by new technologies is high. One thing is clear, however: the best model for the future will combine these two methods, applying them according to the type of product, component, or service the manufacturer wants to purchase.

3.9 TOYOTA: AN ENDURANCE RUNNER

Across the automotive industry as a whole, Toyota's endurance has long been legendary. "Toyota's march to the top is one of the most remarkable examples ever of managing for the long term," commented the *Harvard Business Review* in 2007.[23] This planning far into the future has remained one of the major differences between the Japanese carmaker and the more typically quarterly Anglo-American business planning, heavily oriented toward shareholders, short-term results, and speedy successes. Toyota stands in stark contrast to this, with extremely conservative and thoroughly risk-averse financial planning. The high cash flows Toyota enjoyed for so many years through to 2008 were mainly been reinvested; usually only a relatively small amount was ever redistributed to shareholders: during the period of expansive growth, dividends increased at a lower rate than profits, with the record fiscal years ending 2007 and 2008 being exceptions. Dividends were raised strongly, only to sink back to their 2004 level in fiscal 2010 due to the crisis.

At Toyota, managing for the long term means systematically investing capital in order to leverage scale effects and avoiding spending money on cars in unprofitable niches. In the long term, this strategy provided the company with an enviably high amount of capital reserves. For a long time, Toyota was less dependent on external financial resources and credits than its competitors, and the reserves it had were instrumental in allowing it to weather the storms of the last few years. Even now, Toyota City likes to keep the influence of external providers of capital as low as possible. The priority has always been constant, organic growth with a high level of stability, and that is what Toyota shareholders have always expected, too.

Setting up this system based on profitability and constant capital investment took much time and no small quantity of resources, with Toyota constantly adding to its increasing capital invested in plant and equipment, for example. This strategy, however, weighed heavily on the balance sheet when sales went down in 2008 (see Section 3.1 – The "Toyota shock"). Due to the size of the losses, even market leader Toyota was forced to rethink its strategy of investing in infrastructure and make room for shorter-term profits and higher capital efficiency.

When observers talk about Toyota's "uncomplaining capital," they tend to mean the extraordinary ownership structure behind the organization, a structure which has not only encouraged long-term planning, but made it possible in the first place. One of the biggest shareholders is the company itself: the Company Treasury Stock holds 9 percent of Toyota shares,[24] with other large shareholders being the Japan Trustee Services Bank at 9.97 percent, and the Master Trust Bank of Japan at 5.56 percent (the biggest shareholder in which is Nippon Life Insurance, which is in turn the fourth largest shareholder of TMC with almost 4 percent). These two large pension funds manage most of the TMC pension deposits and therefore naturally have a vital interest in the long-term stability of the company. The second-most important shareholder in TMC is Toyota Industries at 6.25 percent; conversely, TMC has a more than 20 percent stake in Toyota Industries, making it the biggest shareholder in this supplier. TMC is also the biggest shareholder in the remaining members of the Group. It becomes clear that TMC and its companies form a block involving always the same shareholders, which mirrors the jungle of connections between Toyota and its suppliers that we explored earlier in this chapter (Table 3.1).

TABLE 3.1 **Major shareholders (Top 10 Largest Shareholders) (as of March 31, 2011)[25]**

Names	Number of shares held (thousands of shares)
Japan Trustee Services Bank, Ltd.	343,704
Toyota Industries Corporation	215,640
The Master Trust Bank of Japan, Ltd.	191,724
Nippon Life Insurance Company	130,057
State Street Bank and Trust Company	110,672
The Bank of New York Mellon as Depositary Bank for Depositary Receipt Holders	85,886
Trust & Custody Services Bank, Ltd.	84,184
Tokyo Marine & Nichido Fire Insurance Co., Ltd.	67,095
Mitsui Sumitomo Insurance Co, Ltd.	65,166
Denso Corporation	58,678

Notes:
1. The Bank of New York Mellon as Depositary Bank for Depositary Receipt Holders is the nominee of the Bank of New York Mellon, which is the Depositary for holders of TMC's American Depositary Receipts (ADRs).
2. The percentage of shareholding is calculated after deducting the number of shares of treasury stock (312,001,000 shares) from the total number of shares issued.

With this tight network of mutually dependent shareholders, the Toyota ownership fortress is essentially unconquerable: the close weave of stakes, supplier relationships, and business interests makes for a thick and protective barrier encompassing the company, and this network as it stands today was in no small degree a direct consequence of the Asian crisis at the end of the 1990s. During this period, the interlocking ownership system, based substantially on the ailing banking sector, came under serious danger of failing, with the defensive wall of banks slowly crumbling under the pressure of the Asian financial crisis. In Toyota City, the idea of losing equity is, put simply, the worst-case scenario, and so the carmaker rebuilt its castle walls using pension funds and large Japanese insurance companies as bulwarks, marginalizing financial institutions – even though this strategy caused discomfort in the company at the time.

The advantage of this "uncomplaining capital" is that Toyota was not under so much pressure to produce short-term profits or to pay high dividends, and it is the philosophy of Toyota's owners that all parties involved are best served by growth and by increases in productivity. The insurance and pension funds that have the largest shares in the company – along with the Toyoda family with their estimated 2 percent stake – want Toyota to work sustainably, and this is a precondition for the tenacity with which the carmaker prepares market entries or develops alternative powertrain technologies. The first time that shareholders' pressure on Toyota – and on its CEO, respectively – actually became obvious was in the wake of the recall crisis: Toyota shares dropped dramatically and Akio Toyoda at first refused to testify before the US Congress, which sent Toyota shares further downhill. Pressure on him mounted, and after he announced his intention to testify, shares recovered from JP¥3537–3799 (US$38–40.8 respectively; €26.4–28.4).

Just as Toyota wishes to avoid high volatility in its share price – either up- or downward – the company and its shareholders also take a very dim view of large-scale, perhaps risky takeovers, despite any rumors that might spread every now and then about alleged intentions regarding the acquisition of German premium manufacturers.

During Toyota's early years under the management of the Toyoda family, a certain mythology arose, relying on the deep-seated principle that the company wins its battles on its own. Even today, this dogma about not relying on help from outsiders is part of the genetic code and the reflexes of Toyota, and the trio of crises in the last few years have done little to change this. A key event at the heart of these legends was the development of a compact car in the early 1950s. At this stage, the Japanese car industry was still in its very early years, with vehicle manufacturers in the Land of the Rising Sun having more experience with trucks than anything else. Conventional wisdom was that a technology transfer from the USA would be indispensable if the

Japanese wanted to enter the compact segment successfully. Toyota, however, built its first compact car, the Crown, without any form of Western knowledge transfer – and was extremely successful with the model on the Japanese market. This myth still shapes Toyota's determination to manage whatever challenges come up on its own.

What is more, Toyota would be very wary of watering down or possibly even having to rethink its peculiar corporate culture by tying itself up with a foreign manufacturer. Toyota has already seen the problems that can arise when opening factories in Europe and trying to implement the Toyota culture there, and it is well aware of just how tricky it can be to navigate the waters that separate Western and Far Eastern attitudes and expectations. Toyota's rival Volkswagen, for example, has yet to prove that its link-up with Suzuki will work well.

Meanwhile, Toyota joint ventures such as those in Fremont (GM and Toyota) and Kolin (PSA and Toyota) have worked well precisely because production is entirely set up along Toyota lines; mutually agreed to by both joint venture partners. Nevertheless, these working relationships are not set in stone, as the cancellation of the GM partnership proves, and Peugeot, too, may be moving away from the Japanese company, having announced in early 2010 that it wanted to rethink its cooperation strategy and made very public advances to Mitsubishi. Peugeot is already producing a four-wheel-drive and the i-Miev electric car with the Japanese corporation, which are sold under the names Ion and Citroën C-Zero by the French group. The two companies have been jointly running a plant in Kaluga, Russia since April 2010. Moreover, the two organizations are trying to tie up capital, and despite the last-minute cancellation of an exchange of shares, these plans have not yet been abandoned entirely. For quite some time now, the French and the Japanese have been reported to be in talks about an entry-level vehicle for emerging markets, and this Peugeot-Mitsubishi car might well share parts with a future version of the Peugeot 107, which is currently being built with Toyota at the Kolin plant in the Czech Republic. This situation looks as if it could lead to Peugeot dropping one partner for another, and the partner going home alone from the party might even be Toyota.

Nevertheless, even if Toyota loses joint venture partners and struggles with excess capacity, the Japanese company is very unlikely to use its capital to buy a rival carmaker or enter into a capital swap like Nissan and Renault. Even in tough times, Toyota's self-image as a company requires it to use its resources to try and bring itself back on a track to growth under its own steam and relying on its legendary endurance.

The US$50 million (€35 million) stake Toyota took in Tesla, however, is a prime example of how Toyota is willing to make smaller investments whenever this suits its long-term strategy and fills a gap in its own capabilities. The capital injection into Tesla made it possible to reanimate Nummi, and also gave Toyota a roughly 3 percent share in the Silicon Valley carmaker just before it went to the stock market. Toyota will also profit from Tesla's e-car technology, while the American company is said to have built-up losses of US$230 million (€160 million). In 2010, Tesla even reported a third-quarter loss of US$34.9 million (€24 million) compared with a loss of US$4.6 million (€3.2 million) in the same period of the previous year when the company had still been privately held. Tesla clearly had been on the lookout for an investor with deep pockets to join its partner Daimler, which already holds a 10 percent stake. Toyota and Tesla was therefore something of a match made in heaven.

Wherever it makes strategic sense, Toyota tries to cooperate – for example, where alternative powertrain systems, battery cells, or automotive platforms are concerned. Toyota and Daimler, for example, are said to have been in talks about cooperation in developing platforms for the upcoming A and B Class Mercedes. Eventually, the German luxury car manufacturer, however, decided to sign a deal with Renault. Meanwhile, Toyota has already delivered a compact-car platform to Aston Martin, which is basing the Cygnet on it, a sporty city vehicle. This kind of cooperation allows Toyota to put its efficiencies of scale to good effect and see spare capacity put to use. Partners like Aston Martin, on the other hand, are able to bridge a strategic gap by buying Toyota platforms that would otherwise have been costly. Consolidation in the industry, which was only accelerated by the crisis and led to tie-ups like Fiat/Chrysler, VW/Suzuki, and Daimler/Renault-Nissan, will, however, almost certainly wash over Toyota and leave it untouched, even if the Japanese company is now seriously weakened by continued financial issues resulting from the turbulence of the last few years. Consolidation has always been seen with suspicion in Toyota City, where links such as Renault-Nissan are closely watched.

Toyota has proved again and again that it can work its way out of crises, even the most punishing ones – such as the big recall disaster in its most important market, the USA, or the chaotic aftermath of a once-in-a-century natural disaster. By the end of 2010, Toyota already seemed to have overcome the worst of the economic and recall crises, posting acceptable profits again and having come to terms with the handling of recalls; by the end of 2011, it is more than likely that Toyota will be well on its way to recovering from the effects of the earthquake. The long-lasting effects of these crises on Toyota's psyche, however, might not be fully obvious yet. Above all, the recalls of 2009–10 in the USA have had the greatest effects on Toyota. Even if Toyota's own overcapacity and production methods made it more vulnerable to the fall-off in sales in 2008–09 than others, the economic crisis and the 2011 earthquake were external events which caught the manufacturer unawares; but the quality crisis and the ensuing public relations disaster have their roots nowhere else but in the company.

3.10 THE DISASTER: "RUNAWAY TOYOTAS"

In August 2009, just south of San Diego, a 911 operator receives a panicky call from a Lexus driver: "Our accelerator is stuck ... we're in trouble, there's no brakes." The connection breaks off, then a second call comes in, this time from one of the three terrified passengers in the back seat of the Lexus; the police officers on 911 duties record the call: "We're approaching the intersection...Hold on guys, pray, pray!" Then the Lexus crashes; all four people in the Lexus are killed.

This "runaway Toyota" incident sparked the biggest-ever recall for the automaker – and its biggest-ever PR disaster. In Toyota's biggest, most important sales region, a country whose roads are used by millions of Toyota and Lexus vehicles every day, the company found itself fighting a severe loss of trust and its reputation for quality. The timing could not have been worse, either, with the US market just starting to recover from the punishing effects of the economic crisis – and under usual circumstances, growth in the American car market would have meant handsome growth for Toyota as well (Figure 3.3).

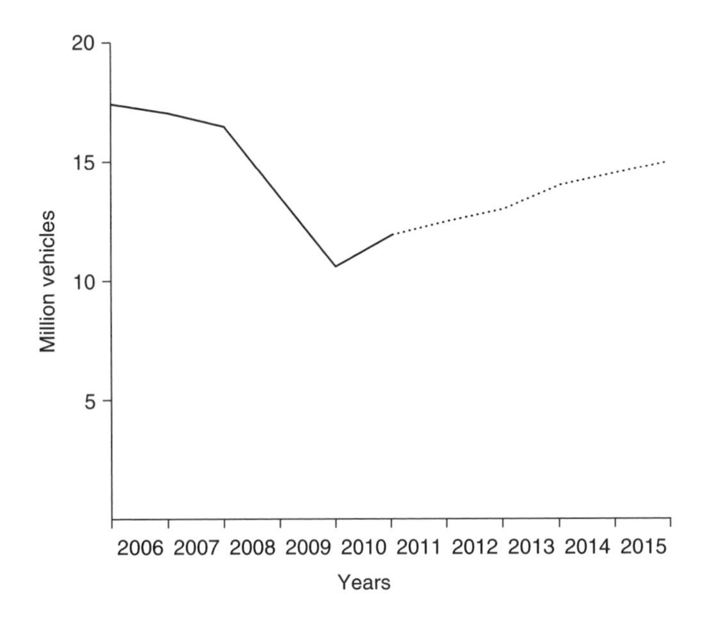

FIGURE 3.3 **Sales forecast for the USA until 2015 (all vehicle segments including commercial vehicles).**

Source: 2005–8 figures taken from *Automotive News Europe* (Global Market Data Books) and MarkLines; modeling by PA Consulting.

The Japanese carmaker was caught completely off guard by the "runaway Toyota" crisis, however. Through September 2009, it became clear that the crash in San Diego was not going to be a one-off tragic story; the National Highway Traffic Safety Administration (NHTSA) began investigating the Lexus accident in light of the fact that the actual cause could not be ascertained. Before any announcements were made, Toyota recalled 3.8 million cars to have their floor mats exchanged for new ones, explaining that the "unintended acceleration" had come from "sticky floor mats." Even at that time, this step was the biggest-ever vehicle recall for the quality-obsessed Japanese, but the media were speculating that there was probably another cause for the "runaway Toyota" that had nothing to do with mats.

By early January 2010, Toyota was under pressure – not just from the press, but from the NHTSA too – and on the 21st of that month it was forced to start a second recall of over 2.3 million vehicles. The reason this time was not sticky mats, but sticking gas pedals. Just five days later (on January 26) Toyota stopped the production of eight models – half of the US range – and prevented dealers from selling them. If it was not one already, at this moment the quality disaster morphed into a disaster of epic dimensions. Toyota's reactions were poor, giving the impression that it had simply not expected the storm of outrage among customers, dealers, and the media. The company was almost completely silent during this crucial period, leaving dealers, the competition, media experts, and car engineers to speculate about the cause of the sticking gas pedals for weeks on end. Toyota limited itself to saying that it intended to repair the gas pedals as quickly as possible.

Instead of a public apology by CEO Toyoda, who remained aloof and distinctly camera-shy, American television viewers saw Ford CEO Alan Mulally talking about competitors' sticking pedals while stressing the high safety standards of Ford cars. Instead of talking to Toyota managers, television journalists went to dealers – who were themselves sinking under the weight of media attention – to ask them what should be done. The dealers tried to calm the situation, stressing that only a very small number of Toyota vehicles would have accelerators that could get stuck. This was, of course, nowhere near adequate to calm the unsettled customers.

Since the fall of 2008, Americans had gotten used to relentless crises: the Lehman Brothers collapse, the near-death experience of the entire world economic system, the decline and fall of the American Big Three car manufacturers – all accompanied by unparalleled big-government rescue packages – and now, the Runaway Toyotas. The crisis for the Japanese carmaker started spiraling in on itself until it had developed a dynamic and dramatic structure of its own, and yet Toyota did nothing to try to take control of its communications. The media just could not drop the idea that people who had bought Toyota models precisely because of the carmaker's good reputation for quality and safety were now at risk of death due to technical malfunctions; this led to a wave of "Toyota-bashing." On all networks on televisions screens from coast to coast, people from every walk of life shared their horror with America: pregnant women talked about how scared they were to get into a Toyota vehicle; honest, hardworking commuters gave reporters statements like this one: "I put my life in danger every day to get to work to make a living for myself to pay my bills, including this car," said one driver, showing just how injured American consumers were feeling in the wake of the shocking revelations.

A deep-seated feeling of constantly being betrayed by big business of whatever sort and of being lied to by their own government had been festering in consumers' hearts for a long time, and the Runaway Toyota crisis triggered the outpouring of a whole wave of resentment. The NHTSA and Secretary of Transportation Ray LaHood, however, felt the fire of the American public, too, as the investigation into "unintended acceleration" dragged on. In the absence of results, many Americans were quick to accuse Toyota of greed and of putting lives at risk in order to grow fast in the US market. Again and again public discussion focused on whether electronic or software problems had been the true cause of Toyota's instances of sudden acceleration – if so, the changing of floor mats and repairing gas pedals would have been useless.

For Toyota, a company where a public loss of face is to be studiously avoided, this kind of media war was a humiliation of unthinkable dimensions. Attempts to blame the American supplier CTS were nothing more than a helpless attempt to find the ultimate cause of the disaster, since the image crisis was getting entirely out of hand. Accelerators may well be produced by suppliers, but they are manufactured according to tight order specifications, and the OEM then has the task of checking incoming parts to make sure that they meet the requirements. Only suppliers who are fully integrated into the production system, unlike CTS, can have any effect at all on the final assembly of the car, at which stage problems might be caused by the interaction between various parts even where the parts delivered completely match the specification.

After this ill-advised tactic, the situation for Toyota went from bad to worse. On January 29 on the sidelines of the Davos World Economic Forum, Akio Toyoda spoke publicly for the first time about the recalls: "I am very sorry that we are making our customers feel concerned. People can feel safe driving in the current situation. Please trust that we are responding so it will be even safer." By now, however, this was too little, too late.

The next stage of the media war followed, and things took a slightly surreal turn. On February 3, Secretary of Transportation LaHood told Toyota drivers to leave their vehicles on their driveways until they had been repaired; and although he later retracted this comment, the damage had already been done. Meanwhile, a whistle-blower passed documents to the authorities that appeared to be intended for Toyota's internal use only and which, so it was claimed, proved that Toyota had known about the quality issues all along and had tried to cover them up. Although the claims were later discredited to quite some degree, the damage to Toyota's public image was already done.[26]

Moreover, the stream of bad technical news just did not stop, despite Toyota's engineers having found two solutions for the gas pedal problems. A thin metal sheet – called a "shim" – would be inserted in the accelerator mechanism to increase tension in the accelerator and help prevent it from sticking. The automaker explained that wear and tear could lead to increased friction within the mechanism, slowing reaction and, occasionally, causing accelerators to stick. The repairs, detailed Toyota, were being made to the contact point for the springs which push the pedal back up to its starting position; the shim was being added to make the pedal springier and to protect the mechanism from further wear and tear. Meanwhile, CTS was racing to step up production of a redesigned gas pedal that was intended to fix another sticking issue, which was believed to be connected to condensation inside the pedal.

Unfortunately, mere days after this announcement the Toyota flagship vehicle Prius was also recalled to deal with additional complications, which in some cases could have led to a delay in the brakes acting. Additionally, Toyota was forced to expand its recalls of various Toyota and Lexus models to Europe and China. More bad news came in the form of subpoenas for class-action lawsuits against Toyota. Hundreds of Toyota drivers were suing the company for potential loss in value of their vehicles on the second-hand market, and the high damage demands characteristic of American class actions – not to mention the lawyers' costs, too – aggravated the situation. By this stage of the recalls debacle, criticism of Toyota's crisis management – or lack of it – was piling up. Toyota seemed simply incapable of communicating that it was doing everything possible to try and solve the issue at hand, and although Congress was setting up emergency committees to examine the problems, Toyota still had not undertaken any press or public relations initiatives of note.

It was no surprise that the competition – GM above all, but also the European rival Volkswagen – did not wait to start turning Toyota's crisis to their advantage. GM offered Toyota drivers a premium of up to US$1000 if they wanted to exchange their Japanese car for a GM model. Ford followed suit, and even Volkswagen was seen strengthening its sales activities, running an extra advertising campaign to persuade drivers looking to change from Toyota to purchase a VW. This kind of activity took place against a background of ever-rising mortality figures from past road accidents that the American authorities came to attribute to unintended acceleration in Toyota or Lexus vehicles.

Not until the recalls had extended to China and Europe did Toyoda finally arrange a press conference and publicly apologize for the recalls of Toyota and Lexus models. This would not be the first public apology: after the recalls had started to scale down, Toyota started to change its communications strategy, slowly but surely redrawing the front lines and driving the PR war away from the public eye in order to avoid any further damage to its already ravaged reputation. In early summer 2010, the recalls had passed their peak and Toyota was able to report that more than 70 percent of the defective vehicles had been repaired; and while the expected collapse in Toyota sales appeared moderate on paper, at 114,000 fewer vehicles in fiscal 2010 than in the previous year, the previous year had already seen a huge drop in sales due to the financial crisis. Nevertheless, Toyota's market share in 2010 was another 1.8 percent down on the already poor 2009, languishing at 15.2 percent. Toyota itself booked financial losses related to the recalls crisis of at least JP¥180 billion (US$1.9 billion, €1.3 billion).

Finally, in early February 2011, a ten-month study carried out by Transportation officials, and assisted by engineers with NASA, concluded that electronic flaws were not to blame for the reports of sudden, unintended acceleration that had led to the recall of thousands of Toyota vehicles. Some of the acceleration cases could have been caused by mechanical defects – sticking accelerator pedals and gas pedals that could become trapped in floor mats – that had been dealt with in the recalls, the government study stated. In some cases, investigators suggested, drivers simply hit the gas when they meant to press the brake. The investigation bolstered Toyota's contentions that electronic gremlins were not to blame and that its series of recalls had directly addressed the safety concerns. Even Transportation Secretary LaHood went on record as saying "We feel that Toyota vehicles are safe to drive."

However, Toyota paid the US government a record US$48.8 million in fines for its handling of three recalls and reacted in terms of car safety as well: since these large-scale recalls, Toyota has installed brake override systems on new vehicles. The systems automatically cut the throttle when the brake and gas pedals are applied at the same time. Furthermore, throughout 2011, Toyota has been running a "better safe than sorry" strategy, recalling several smaller batches of vehicles for small-scale technical defects. The company also created engineering teams to examine vehicles that are the subject of consumer complaints and appointed Rodney Slater, a former US Secretary of Transportation, as chief quality officer for a North American Quality Advisory Panel to tackle criticism that its US division did not play a large enough role in making safety decisions.

Toyota's Recall Disaster – A Chronology

August 28, 2009

A Lexus ES 350 crashes with four people on board near San Diego; before the fatal collision, the driver is able to inform the police that his accelerator is stuck.

September 29, 2009
Toyota recalls 3.8 million vehicles to auto shops across the USA. It explains that problems with sticking gas pedals may be to do with floor mats.

November 25, 2009
Further measures to combat the floor mat issues are announced, including a reconfiguration of the acceleration system, new mats, and a brake override system.

January 21, 2010
Toyota recalls 2.3 million vehicles; it explains that problems other than the floor mats might be causing accelerators to stick.

January 26, 2010
Under pressure from the National Highway Traffic Safety Administration (NHTSA), Toyota halts production of eight models in the USA.

January 27, 2010
Toyota extends its recalls of cars with possible pedal problems by another 1.1 million vehicles. Now former joint-venture partner General Motors is involved, with Pontiac Vibes produced in Nummi (Fremont, California) that were partly built with Toyota components needing to be repaired, too. The problematic gas pedals had been built into around 99,000 cars assembled in Nummi in 2009 and 2010.

At the same time, competitors start to react to the recalls. General Motors is at the vanguard, encouraging unsettled drivers to swap from Toyota to one of its models.

January 28, 2010
Recalls reach China as Toyota is forced to examine 75,000 RAV4 vehicles due to acceleration problems.

January 29, 2010
Recalls are extended to Europe, where Toyota checks 1.8 million cars for possible sticking gas pedals. Eight models are affected, as is joint-venture partner PSA Peugeot Citroën: both the Peugeot 107 and the Citroën C1 are identical to the Toyota Aygo. The companies have to recall 100,000 107s and C1s produced at the Czech plant at Kolin.

At the World Economic Forum in Davos, for the first time CEO Akio Toyoda apologizes briefly for the recalls, on Japanese television.

February 1, 2010
Toyota goes public with its solution for the accelerator issue; it starts to send out the new parts immediately.

February 2, 2010
The US Secretary of Transportation Ray LaHood criticizes Toyota's crisis management in no uncertain terms and goes on record to Associated Press to term Toyota "a little safety-deaf."

February 3, 2010
LaHood warns American Toyota drivers not to use the vehicles until the recall repairs have been carried out; he retracts the warning soon afterward.

February 5, 2010
At a press conference, Toyoda offers a full public apology for the recalls.

February 9, 2010
Toyota is forced to recall another 437,000 vehicles worldwide due to braking problems; the flagship vehicle Prius is among them.

February 12, 2010
Due to potential problems with the front driveshaft, Toyota recalls 8000 Tacomas.

February 24, 2010
Toyoda is called as a witness before the US Congress.

March 1, 2010
Toyoda travels to Beijing to apologize for the recall of the RAV4.

March 8, 2010
At a meeting with Toyoda, former Japanese Prime Minister Yukio Hatoyama demands better quality control at Toyota in order to regain customer confidence. The technical defects, says the premier, would not just have damaged the carmaker's image, but also Japan's reputation as a high-tech economy.

March 23, 2010
On a hastily arranged trip to Europe, Toyoda meets workers, suppliers, and dealers at Toyota's European HQ in Brussels, thanking them for their efforts during the recall crisis.

March 30, 2010
In addition to its recall efforts, a new global quality committee to coordinate defect analysis and future recall announcements is announced by Toyota along with a Swift Market Analysis Response Team ("SMART") in the USA to conduct on-site vehicle inspections, expanded Event Data Recorder usage and readers, third-party quality consultation, and increased driver safety education initiatives. Industry analysts note that the parts might not have been recalled owing to factory errors or quality control problems, but rather because of design issues leading to consumer complaints. As a result, the global quality committee aims to be more responsive to consumer concerns.

May 10, 2010
LaHood travels to Toyota's central HQ in Japan to talk to Toyoda about improvements in quality assurance.

May 18, 2010
Toyota pays a record fine of US$16.4 million in order to bring legal wrangling with US authorities to a close; nevertheless, it does not accept responsibility for the accidents.

May 21, 2010
Toyota recalls 3800 Lexus LS vehicles due to steering problems.

May 27, 2010
The NHTSA reports that the number of deaths resulting from sticking gas pedals in Toyota and Lexus vehicles was could be higher than it had previously assumed. According to NHTSA, the authorities had received reports of about 71 serious accidents potentially resulting from unintended acceleration; 89 people had lost their lives and 57 were injured in the accidents concerned.

July 5, 2010
270,000 Crown and Lexus models are recalled worldwide to rectify potential production issues with valve springs.

July 29, 2010
412,000 Avalons and LX 470s are recalled in the USA for replacement of steering column components.

August 11, 2010
The Wall Street Journal reports that experts at NHTSA had examined the "black boxes" of 58 vehicles involved in sudden-acceleration reports. The study found that, in 35 of the cases, the brakes weren't applied at the time

of the crash. In nine other cases in the same study, the brakes were used only at the last moment before impact.

August 28, 2010
About 1.13 million Corolla and Corolla Matrix vehicles produced between 2005 and 2008 are recalled in the USA and Canada for Engine Control Modules (ECM) that may have been improperly manufactured.

October 21, 2010
Toyota announces three separate recalls of a combined 1.5 million vehicles, including 740,000 in the USA, over concerns that brake-fluid leaks could eventually hinder brake performance. The recall involves the models Avalon, non-hybrid Highlanders, and several Lexus models. Another 600,898 vehicles are recalled in Japan and about 200,000 in China.

December 21, 2010
Toyota agrees to pay two more safety fines for a total of US$32.4 million, the maximum allowable under US law, to settle federal investigations into whether the company had notified regulators of safety defects in a timely fashion. The fines, in addition to an earlier US$16.4 million levy paid to settle a similar probe, brings to US$48.8 million the amount that Toyota has agreed to pay in civil penalties in 2010.

February 25, 2011
Urged by the US Department of Transportation, Toyota has to recall 2.17 million cars, addressing a problem with accelerator pedals getting jammed under floor mats or driver's side carpeting. This comes just days after a joint report by NHTSA and NASA rules out an electronic problem causing sudden acceleration in Toyota vehicles.

March 7, 2011
Following the report of February 8, Toyota also has to recall thousands of SUVs and pickup trucks. According to the NHTSA, some vehicles produced between 2008 and 2011 (such as the Land Cruiser) may have faulty tire deflation monitoring systems.

June 29, 2011
Smaller-scale recalls rumble on. Having recalled 100,000 of its 2001–2003 flagship Prius fleet worldwide at the beginning of the month for poorly wearing steering wheels, Toyota also has to recall 45,000 Highlander hybrid and 37,000 Lexus RX 400h vehicles sold in USA due to faulty transistors in control boards. It seems that these components can suffer from heat during high-load driving.

Another of Toyota's woes is the series of unpleasant class-action suits following the recalls that, at first, threatened to drag on and lead to high costs: legal experts at first estimated possible damages at more than US$10 billion (€7 billion). Nonetheless, by the summer of 2010 it was already becoming clear that the plaintiffs, as well as the media and participants in the public discourse in general, had overreacted in some cases and that damages were likely to be much lower; the findings of the joint NHTSA/NASA report of early February 2011 have of course served to strengthen the manufacturer's position. Nevertheless, even if the damages resulting from litigation do not turn out to be as high as first expected, it will certainly drag on and cause high legal costs: in July 2011, the first 'bellwether' case – which will be used as a benchmark for future cases – was scheduled for trial in 2013.

Meanwhile, Toyota has been doing its best to keep customers happy by offering big incentives along with each new Toyota or Lexus sold. A recovering US market had also helped prop up Toyota's sales going into 2010, with Toyota reporting sales of just over 27 million vehicles in the USA from April 1 2009 to March 31 2010. Despite recovering sales, however, the gas pedal scandal had woken some rather unpleasant memories about earlier problems in the US market: the so-called "compact car wars." This trade dispute took place in the 1990s and shared certain characteristics with the "Toyota-bashing" of 2010, namely the failure of the American Big Three to defend themselves against Japanese carmakers in their own heartland. In the 1990s it was Toyota, Honda, and Nissan that, having been established in the US market since the 1970s, were continuously increasing market share with imported vehicles at the expense of domestic manufacturers, which decided not to give up their home market without a fight.

Taking a Hammer to Toyota

The impetus for Toyota to move production out of Japan often came from protectionist measures in export markets, and this was nowhere more the case than in the USA. As long as Toyota had no plants in the country, it was often the victim of anti-Japanese sentiment, with their successful export models produced in Japan and sold in the USA seen as a danger to American jobs. Members of the United Auto Workers union are even said to have vandalized random Toyota vehicles with hammers at the height of the US–Japan trade disputes. In order to reduce tensions, Honda, Nissan, and Toyota began to build factories in the USA, and the Big Three were suddenly faced with the fact that they had not rid their home market of the Japanese competitors, but were now going head-to-head with them there. Toyota, meanwhile, feared the power of the UAW and built its plants in Southern states where the UAW's influence is less strong.

Starting in 1988, Toyota began building factories on its own in the USA, and from then on North America became its principal growth market; as was true of Toyota's Japanese rivals Honda and Nissan. Toyota grew in the market, and grew with it, too. Meanwhile, the domestic manufacturers – the Big Three – found themselves increasingly under pressure and could only keep up sales through aggressive price cuts as the three Japanese manufacturers won over customers with attractive products at comparatively good value. Furthermore, with its preference for automatics and speed limitations, the US market is far closer to the Japanese understanding of the industry than the European countries, which are characterized by fast highways, manual transmissions, and a high demand for diesel engines.

Toyota increased its productivity in America year after year and, on the basis of an optimized cost structure, began to gain market share at the expense of the established manufacturers. Moreover, it developed considerable skill in handling the specific aspects of the US dealer system: Toyota allowed car sellers to earn very high margins, and the carmaker remained loyal to its dealers, encouraging their growth rather than forcing them to compete with each other through a high density of dealerships as some other manufacturers are known to do. In return, the sales network remained highly loyal, well motivated, and produced excellent results. In a testament to the strength of this cooperation, it has by and large stayed with Toyota through thick and thin – despite the long periods of uncertainty and the workload resulting from calming customers down and repairing their vehicles. Truly, Toyota's success in the USA was always partly its success with dealers.

By the mid-1990s, the American manufacturers had had enough of this success though, and used a Japanese-American automotive conference to pressure the Japanese into accepting voluntary export limits. In order to ease tensions, Japan also agreed to sell GM vehicles through home dealerships on the Japanese islands and to source fixed quantities of parts from American suppliers. These voluntary export limits only lasted for a few years, however, and despite the attempts of the American Big Three, the Japanese continued to grow their market share in the USA. Even today, it is not uncommon to hear the Japanese manufacturers being blamed for the decline of the American car industry, and the Far Eastern carmakers are as anxious as ever to avoid these kinds of allegation. The trade disputes of the 1990s have left deep psychological marks on Nissan, Honda, and Toyota, and the three therefore try to maintain as good a relationship with the US authorities as possible. Toyota, for example, has in the past largely refrained from aggressive discounting so that it can avoid the accusation of having grown to be number one by applying dumping strategies to the American market. There are rumors that Toyota once even increased prices by 1 percent across its entire range in order to improve the competitive environment for its rivals. Whether this is true or not, it is certainly the case in

America that Japanese firms have historically been anxious to avoid stirring up trade conflicts, with anything that may lead to "trade friction" (as it is known internally at Toyota) being a complete taboo; this is quite probably the reason why Toyota did not hesitate to accept the NHTSA fine in April 2010.

However, as with the voluntary export limits in the 90s, Toyota's restraint from discounting is just that: voluntary. The manufacturer has been severely wrong-footed in the US market, and the combined effects of the ongoing recalls, the natural disaster and unexpectedly resurgent domestic competition in the form of GM and Ford, have seen its sales there shrink back to 1.9 million for the year ended on March 31 2011. This is an even worse performance than the year to March 31 2009, which encompassed the two most punishing quarters of the financial crisis. Yet the company still sees the North American market as very much its own back-yard and, after the turbulence of 2011 is past, may well try to buy back some of the market share it has lost since 2010 – ironically by using their favored tactic of "pile 'em high, sell 'em cheap." Just following the half-year figures for calendar 2011, Toyota announced that it would not only return to full production as planned, but would produce an extra 350,000 vehicles in Japan between October 2011 and March 2012. This will allow the company both to support Japanese efforts to recover lost economic growth and to flood its US market with extra production, probably accompanied by some attractive incentives.

Beyond its sheer size and importance to the company, there is a reason that Toyota sees America as worth fighting for. Just as its major rival VW looks back at decades of experience with the peculiarities of the Chinese car market, Toyota has amassed considerable knowledge of American car buyers and what makes them tick. J.D. Power studies on complaints among car drivers and the reasons for them had no qualms about using Toyota as the measure for other manufacturers, and while VW had trouble in the USA because it had not understood that customers there are not willing to pay for technical refinements that are not immediately visible, Toyota had understood exactly what Americans wanted. Offering nothing more than a handful of simple, uncomplicated models – including the best-selling Corolla – the Japanese oriented themselves more toward service and were very successful. The world's best-selling car, the Corolla, may seem simple in comparison with the VW Golf, for example, which is a complicated, compact car pushing at premium standards in some areas and therefore has difficulty winning over the price-conscious American car buyer.

Meanwhile, Toyota has always tried to anticipate its customers' needs, whether in terms of its range of models or its hybrid technology, and has adapted itself perfectly to American tastes, and until the recalls, this was clear from J.D. Power customer satisfaction studies. J.D. Power, financed more or less entirely by car companies, measures a variety of factors, including customer satisfaction, vehicle quality, service and after-sales, dealership standards, and brand strength. In its 2011 survey, the effects of the recalls showed (see Figure 3.4), with Toyota dropping from fourth to fifth place between 2009 and 2011.

Then again, the Japanese carmaker will surprise observers more if it does *not* make every effort to pull itself out of this slump. Toyota is sure to learn from its painful mistakes, and it has shown before that it is capable of doing this – especially in the US market, which has been more decisive for Toyota's development and success than any other. In a sign, for example, that the quality panel founded in 2010 has real teeth, its chair Rodney Slater went on

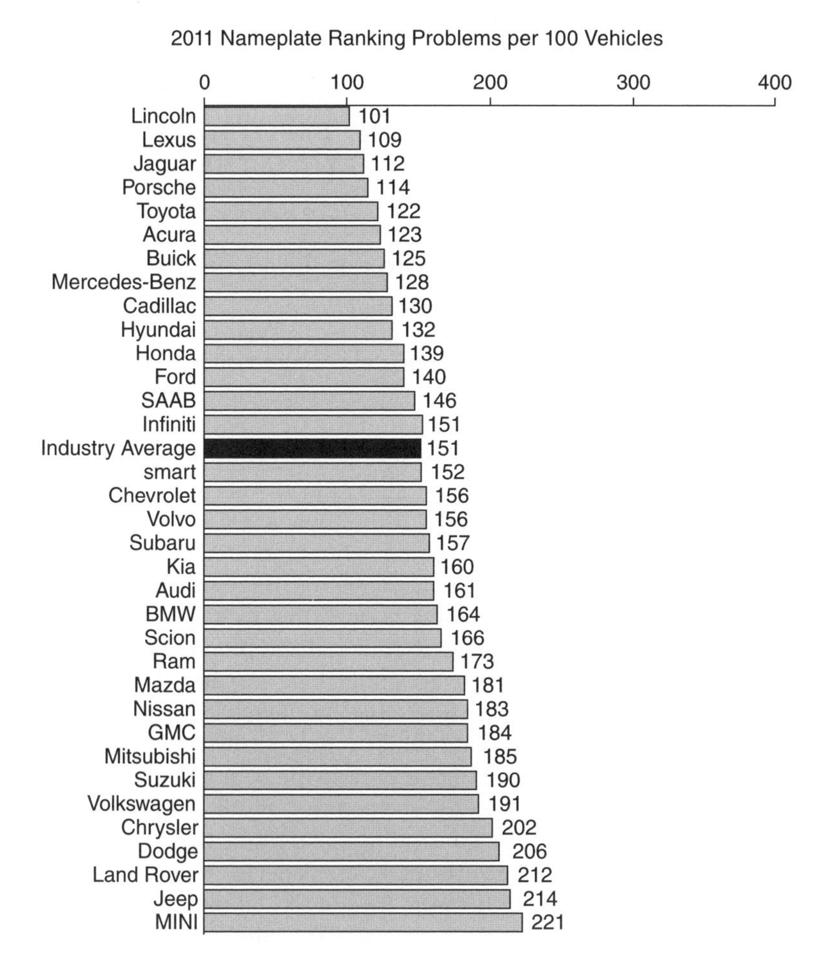

FIGURE 3.4 **J.D. Power and Associates – 2011 vehicle dependability study, US market, released on March 17, 2011. Dependability is measured by the number of problems leading to customer complaints per 100 vehicles.**

Source: J.D. Power and Associates – 2011 Vehicle Dependability Study^{SM}.

record in May 2011 as stating that Toyota "needs to shift the balance somewhat toward greater local authority and control." CEO Akio Toyoda replied to this as follows: "We appreciate the panel's efforts to help us further strengthen our processes, and we thank this distinguished group for their recommendations. Over the past year, Toyota has learned a great deal from listening to the panel's valuable counsel. Their advice has been reflected in the meaningful steps we've taken to give our North American operations more autonomy and become an even more safety-focused and responsive company." For a manufacturer as bent on centralization and control as Toyota has been, this is quite a change in tone.

3.11 CHANGE OF GENERATIONS: EXPLORING FUTURE POSSIBILITIES AND CARRYING THE LOAD OF THE PAST

The recalls did not just have a great effect on Toyota the organization, but also on its CEO, for whom the events – along with the economic crisis –were something of a baptism by fire. Beyond Akio Toyoda, however, the recalls have triggered changes that are becoming ever more visible today.

The reactions of the company a mere two years ago show just how far Toyota has come. Faced with the sudden advent of the worldwide sales crisis in late 2008, at first the carmaking giant had reacted sluggishly enough. When the recalls started, however, the as yet untested Akio Toyoda appeared positively paralyzed. Upon taking office in early 2009, Toyoda, grandson of the almost beatified company founder and patriarch, whose monumental bronze statue is set up at Toyota City HQ, had been the Great White Hope of the company – and of half of Japan, too. His task was no less than to help the company to recover from the economic crisis and to walk tall and proud again. The mythological maxims about achieving everything alone, the Toyota Precepts, and the revolutionary production methods that had helped Toyota survive the early years – the weight of this mythology – was entrusted to Akio Toyoda, and he was supposed to bring all of this back to life, to put the spring back in the step of the giant. The young CEO seemed well aware of the expectations resting on him and, rather than continuing the reforms that his predecessor Watanabe had initiated to help Toyota take quick decisions and make radical changes,[27] at the beginning of his rein Toyoda settled on a "back to basics" strategy, a return to the Toyota precepts and the core values and beliefs of over 70 years of Toyota history.

Toyoda's management style was more like that of a reclusive monarch than the CEO of the world's leading manufacturer. Almost unavailable to the media, he hardly ever appeared in public during the first months; when he did, he often read out a written speech, took as few questions as possible, and then disappeared again. If he offered interviews at all, they contained very little about him – except his love of rally driving. His speeches and answers were generally vague, cryptic, and contained little by way of concrete information, providing journalists covering the sector with a huge interpretative task. In fact, anyone wanting usable information about Toyota's strategic outline was better served listening to the heads of the regional divisions, and Toyoda seemed more than happy to let his subordinates handle the tricky task of specifying strategic goals. The new Toyota chief who had been so hotly anticipated seemed weak, lacking a vision for his company, which was suffering seriously. Even facing the recall disaster, he wanted to avoid appearing before Congress if at all possible, probably in an effort to save the one attribute which he undoubtedly did possess – his untouchability.

By the time he had acquiesced to the public and political pressure and agreed to go to Washington, any leeway he may have had to exercise damage control had disappeared. In the media, he was already considered an indecisive pushover, and the interrogation at the hands of Congress had become nothing more than a pre-scripted humiliation before he had even stepped off of the plane. In addition to this, Toyoda had already used up much of his ammunition with his warning about Toyota's danger of capitulating to irrelevance and death – and this in a company that has a general tendency to wallow in and even enjoy a sense of crisis. Its founders were already warning about the dangers of the self-satisfaction that can develop in boom periods before the company was much over the age of legal majority, believing that Toyota was at its best when it was under pressure. As if to prove the founding fathers right, both the immediate postwar period and the oil crises were times during which Toyota was able to tap previously unsuspected wells of strength, and for this reason, Toyota bosses have always made sure to spread the feeling that the business is in a crisis of some sort. Yet by October 2009, Toyoda had reached a level of scaremongering that even Toyota could no longer handle. Having shot his bolt, he had little rhetoric left to mobilize the real and undeniable atmosphere of crisis that gripped his company as the recall waves reached their height.

The differing expectations placed on Akio Toyoda, who is supposed on the one hand to embody the memory of the much-honored founding fathers and on the other hand to lead the car manufacturer out of the crisis, often seemed to contradict one another. Watanabe had promised that a new company would emerge from the crisis, and Toyoda, despite implementing savings measures, seemed for a long time strangely unable to communicate his vision for a new Toyota, a company that wants to regain its former strength and keep competitors at bay. The quality crisis in the USA did, however, lead to a clear paradigm shift in Toyota City, with Toyota trying to become more transparent – and its CEO more accessible – in the wake of the recalls. Not just in the USA, but in Europe and China too, the Toyota chief went on record to apologize and explain the measures that the manufacturer was taking, and this new willingness to communicate with the press has outlived the worst of the recalls crisis. Toyoda speaks far more regularly with journalists, video recordings of press conferences are posted on the internet, and the organization that was legendary for its ability to stonewall on many issues is now making active efforts to use communications and public relations for its own benefit.

Appearing before the press on May 11, 2010, Toyoda was already able to put his new-found public relations skills to good use, presenting an unexpectedly successful set of figures, with an operating profit of JP¥147.5 billion (US$1.58 billion, €1.1 billion) and a final result of JP¥209.4 billion (US$2.25 billion, €1.6 billion) – despite sales having actually fallen from their level in 2008–9, the fiscal year most affected by the crisis. "I feel that, given these earnings, we are finally standing at the starting line. I believe that this fiscal year marks a truly fresh start for Toyota," he said confidently. Just under a year later, and Toyoda was once again in front of the cameras, this time talking about the effects of the March 2011 earthquake on Toyota, and not missing an opportunity to stress his concern for members of the company and of affiliates who had been affected by the catastrophe. By seeking out opportunities to talk to the international media – such as during his visit to Microsoft in April 2011[28] – Toyoda was able both to show his human side and to reassure investors, analysts and car buyers worldwide that Toyota was dealing well with the aftermath. The Akio Toyoda who so desperately tried to avoid the limelight in 2009 is now able to deal with it and, in some cases, use it to his own ends. Another indication of the reforms being introduced at the top came when Toyoda traveled to the USA to announce the stake in Tesla in May 2011. While his prefabricated speech may have appeared somewhat awkward by comparison to more off-the-cuff speeches by Tesla CEO Elon Musk or former Californian governor Arnold Schwarzenegger, the message was nevertheless loud and clear: Toyota has a new face, a new CEO who is willing to take personal responsibility for the recalls, and who is willing to risk a new start in the USA of all places. Not unimportantly, Toyota was also able to use the Tesla cooperation to raise its profile as an employer in the region – and with jobless rates running at around 12.5 percent in California, this is no peripheral consideration. Toyoda's promise to create 10,000 jobs in building electric vehicles was therefore in all probability as much a politically motivated statement as anything else. Toyota has well-known pedigree as a skilled lobbyist in the USA and it is clearly trying to use its function as an important regional employer to win back some of the many supporters in important political arenas that it lost to the recalls. Last, but certainly not least, the investment in the electric carmaker Tesla is a concrete example of the path Toyota is looking to take into the future, with sheer size for the first time taking a back seat.

Moreover, the chief had something else to say, too: "By working together with a venture business such as Tesla, Toyota would like to learn from Tesla's challenging spirit, quick decision making and flexibility. That is a big part of the reason we decided to partner with Tesla." While this may sound like nothing more than a little casual flattery, it is in reality a big step for the Japanese giant, and a clear message to its workers at home: one of Japan's most traditional companies wants to speed up its slow decision-making processes.

Implementing the new strategy, with the new Global Vision announced by Akio Toyoda in March 2011, the number of members of the Board of Directors was substantially reduced. Together with other organizational changes, this reform was the precondition for Toyota now trying to rationalize its decision-making processes.

Not only has Toyota shown astonishing resilience in the face of the unholy trilogy of powerful shocks between 2008 and 2011, but it is already planning for the future – and by "future," it means the car industry in the next twenty to thirty years, a typical level of forward planning by Japanese standards that may well appear exaggerated to European and American tastes. This is why the Toyota strategy of today – composed of strict shrinkage, even stricter quality control, clear concentration on growth markets, and the complete hybridization of its range – has been complemented by an objective that many observers would term risky: the pioneer of hybrid technology now wants to focus on fuel cells as the next step in alternative powertrain systems.

3.12 ENVIRONMENTAL PROTECTION: TOYOTA'S DEVELOPMENT OF ALTERNATIVE DRIVETRAINS

Social responsibility for alternative energy vehicles

In terms of research into alternative forms of propulsion, Toyota won the world championship year after year. Until the chaos of recent years, there was simply no other carmaker in any country that was working as hard in this area as the company from Toyota City. In 2008 alone, the planet's largest automotive manufacturer obtained 2379 patents relating to alternative powertrain systems, a good sixth of all patents applied for in this field.

This pace of development has not been driven by Toyota alone, with the Japanese state cooperating far more closely with the automotive industry in several fields than in Europe or the USA. Japanese society is marked by a holistic approach to the economy and society, and it is therefore very much part of the remit of a carmaker – especially one which, like Toyota, is subsidized in various forms by public funds – to contribute to the betterment of society at large. In that sense, Toyota's investment in alternative technologies was not just a logical, but an unavoidable strategic decision. Toyota's obligation is to look for sustainable solutions to societal problems, which include the aging population, climate change, dependency on fossil fuels, road traffic safety, and growing gridlock in the major cities (with 30 million inhabitants, Greater Tokyo is one of the world's largest contiguous urbanized areas). Japanese society does not just expect Toyota, the market leader, to do the legwork in developing future concepts of automotive transport, but to be ahead of the pack.

However, it's not just Toyota's research and development in alternative fields which is supported by the state – both directly and indirectly – but also its progress in areas that, at first glance, are not directly related to mobility. Robotics, for example, is also targeted by government support because, in the Land of the Rising Sun, the technologies of tomorrow are high on the list of political priorities. R&D is not the preserve of academics in laboratories, but a major part of high-level political decision-making processes. High-level policy decisions relating to industrial technology are taken in the Ministry of Economy, Trade and Industry (METI), which has the task of supporting new industries and areas of research with the overriding aim of keeping Japan competitive. While R&D is frequently in a state of dramatic flux, the approach taken by METI is in some ways slightly reminiscent of a 1950s Soviet Planning Committee: futuristic technology is divided up into key areas, and market volume for them in the coming years is estimated; on this basis, a detailed action plan is then produced. Examples of what METI considers key areas are fuel cells, high-end digital electrics for consumers, robots, digital content, health, environmental technologies, energy sources, and business services. Financing is also directed to anti-global warming measures, re-use and recycling, and, of course, developing environmental-friendly cars for the next generation. All of these R&D areas are assessed against overall economic considerations and then prioritized by their market volume and trends in the coming years.

The government intervenes at the most important points, both supporting projects and also managing them to a certain degree; the tools include tax policy, discussions and consensus-finding forums, networking, and, most bluntly, the commendably high level of financing for R&D as well as the considerable societal respect accorded to the people and companies behind scientific and technological advances. The most direct forms of government support arise when the state coordinates and finances single projects, and in all these ways, the Japanese state ensures that the country's economy remains competitive on a worldwide scale. Ten percent of large companies with R&D departments receive direct subsidies for their research through local government, and another 20 percent are direct beneficiaries of support from central government.

In addition to this, major Japanese firms such as NEC, Fujitsu, or Hitachi do not concentrate on covering one area, as is common in the West, but work up and down the entire scale in all areas. A positive side-effect of this is that R&D covers many areas at once and can profit from synergies, leading to far better progress than in the West on technological challenges that require effort from all parts of society, such as the move away from transportation fueled by fossils to electro-mobility based on renewable energy sources (see Section 3.12.3). There are negative results, too, with Japan being very susceptible to the NIH ("not invented here") syndrome.

What is clear is that Japanese car manufacturers have oriented themselves in line with METI's prioritization and have accorded great importance to alternative propulsion systems. This means that Toyota's investment in new technologies is more than just a question of courageous decision making, but also a question of fulfilling its obligation to Japanese society as a whole. Western observers keen to deride some areas of Toyota R&D as eccentric tend to forget this political and societal dimension. A classic example was when then CEO Watanabe proudly announced the dawn of a new age of mobility and arrived at a Tokyo Motor Show press conference in the i-Real study, dubbed a "high-tech wheelchair" by some observers. Yet this is simply missing the point of the three-wheeled, one-person, battery-operated vehicle intended "for passenger transport in the urban environment" – conceived primarily for older consumers in cities who are no longer licensed to drive cars but who need to remain mobile. The slim-line electro-mobile is intended to allow Japanese senior citizens to drive on sidewalks, in pedestrian areas, and even into elevators.

Seemingly extravagant research into complicated robot technology is another area that can be explained by orientation toward government priorities. Another Toyota novelty in 2007 was a robot that was able to play Edward Elgar's "Pomp and Circumstance" without fault on a real violin. Beside this pleasant accomplishment, however, the robot would also be able to take care of older people in need of care. Man's best friend as a robot? This is not some utopian R&D pipe dream, but a concrete priority set by the Japanese government. "Now we want to accelerate the development of robots that make a contribution to society, drawing on our knowledge and innovation in the field of automobiles," said CEO Watanabe as he presented the humanoid. In this particular field, competitors like Honda and Sony may well be far ahead of Toyota, but in view of demographic trends, Japan is the world's biggest potential market for robots. More than 30,000 Japanese senior citizens are over 100 years old (out of a total population of 127 million), and this is reason enough for Toyota to try to offer a solution in this area.

The trend in the West in recent years toward "corporate social responsibility," which is now a part of most industry sectors, is often nothing more than greenwashing. In Japan, responsibility for society is nothing less than the ground on which Japanese companies build; without it, society would reject the company, and R&D implemented in real products and business models is therefore a major part of Toyota's justification for its own existence. Toyota never tires of proclaiming its founding motto: "Contributing to society through the manufacture of automobiles."

The hybrid, saving the company

Due in no small part to these differences in the Japanese corporate environment, Toyota started research into alternative energy vehicles early – like many of its Japanese competitors. At first, the overriding aim was to cut fuel consumption for coming generations of cars, and when the first mass-market hybrid was ready in 1997 in the form of the Prius, suddenly the somewhat unlikely Japanese manufacturer was catapulted into the limelight of the world car industry.

The development of a fully hybridized vehicle was not unknown territory for the automotive sector, but only a company with the level of endurance for which Toyota is known could have kept plowing ahead through unresearched areas, with all the costs this first-mover status reliably entails, all the more since car buyers were frequently skeptical of hybrids. Toyota's sales of the first Prius went up gradually, and it was not until August 31, 2009 – 12 years after the first Prius went on sale – that the manufacturer was able to report that it had sold its millionth hybrid vehicle. Now, however, there are more than 3 million Toyota and Lexus hybrids on the world's roads. The Prius, the best-selling hybrid ever produced, cracked its 2 million mark in October 2010.

The flagship hybrid vehicle in fact is so successful in Toyota's major markets Japan and the USA, that Toyota decided to turn "Prius" into an entire brand family: with the Prius V, a family car with a spacious interior and the Prius C concept, unveiled at the Detroit Motor Show, the technology leader matures into a brand with an identity of sorts and a character of its own. The Prius V is supposed to cater for young families, while the C concept is designed for young urban customers.

In accordance with its company philosophy, once Toyota had taken the decision to mount two engines in one vehicle, this ambitious goal was tenaciously – and expensively – pursued to completion. At the same time, thanks to its more than ten years of making hybrid vehicles, the company is now at the undisputed technological head of the field. This lead is being solidified by

Toyota's high-volume production and the discoveries that have resulted from this in the course of the years. If, for example, Toyota can save production costs – roughly estimated – of 20–30 percent per vehicle every time it doubles the volume of hybrids it is selling, then it becomes clear that Toyota will remain able to produce hybrids at far more competitive costs than its main rival Volkswagen for many years to come. Even if VW somehow managed to suddenly produce (and sell) 1 million hybrid vehicles, Toyota would still have a clear cost advantage due to its lead of more than a decade and would remain able to push down costs by another third. All of this is imagining the situation as it would be were Toyota to take its foot off the technological accelerator, but it is by no means doing this and, by 2012 at the latest, will have reached the next level of evolution in hybrid technology: the plug-in. Furthermore, Toyota is now selling licenses for its technology to smaller competitors – e.g. to Mazda – and can now further increase its economies of scale by doing so. Nevertheless, there is a lot of debate in the automotive sector about the profitability of hybrids, even given high-volume, low-cost production. While there is almost no one in the industry who would dispute that Toyota is now in the black with models like the Prius on an operational level, only Toyota itself knows whether the high costs in overall R&D needed to produce the model have yet been recouped through sales.

Toyota's advantage, however, is more than just a matter of costs, and has a lot to do with the performance and reliability of its hybrid systems. The Prius has not ceased to improve from generation to generation (it is now in its third), especially in terms of the components Toyota itself produces, such as power electronics and complex battery management. Even the less complicated parts of the system, such as high-voltage cabling, have been tirelessly tinkered with by Toyota, and while competitors are still forced to use expensive copper cables and even more expensive plug connections, Prius runs on cheap aluminum cables and robust, simple plugs. Another example is the impressively high performance of the Prius electric motor, not to mention its size: the Japanese engineers have succeeded in packing 60 kilowatts of power into a comparatively small space. Meanwhile, this kind of improvement makes the Prius financially and environmentally ever more attractive to car buyers. In July 2011, a trial of the third-generation plug-in Prius in London showed that the newest edition of the vehicle is 27 percent more fuel efficient than an equivalent diesel car. Put in years, observers estimate that Volkswagen will need until 2015 to reach the level of hybrid technology that Toyota has achieved in 2011.

At an international level, the manufacturer was able to establish itself as a "green" company without any expensive marketing campaigns, even in markets where Prius sales remain very low: Germans, for example, rate Toyota as the world's most environmental-friendly brand, despite the fact that Prius sales in Germany, Europe's largest car market, trail at a few thousand vehicles per year. In addition to this, hybrid technology was a good way for Toyota to compensate for its lack of competitiveness in diesel engines. In contrast to many other parts of the world, diesel is big in Europe. In the 1980s, Volkswagen introduced direct-injection diesel technology, and this permitted an increased use of this far more economical fuel. Newer innovations followed fast, with Fiat cooperating with Bosch to produce the common rail injection system. With its engineers having grown up in an economy in which diesel fuel had not played a great role in passenger car business, Toyota completely underestimated the importance of this trend. In the European markets, domestic manufacturers regained the upper hand and market share, and in some countries, up to 75 percent of new cars were fitted with diesel engines. At first, Toyota tried to develop diesel engines for this difficult market, but it is still behind on this front even today, and hybrid technology was in no small measure intended to compensate for this disadvantage. Toyota hoped that the fuel savings offered by hybrid cars compared to gasoline cars would allow it to recapture market share in Europe while continuing to expand in markets with an affinity for hybrid technology such as the USA. As the years went on, Toyota completely integrated the hybrid strategy into its company philosophy, and this became yet another aspect of its worldwide domination.

In the Japanese market, however, it was less a case of rivalry with diesel technology and more one of environmental concern that spurred Toyota on in its development of hybrid technology; and while sales of conventional gasoline vehicles in the Japanese market declined by more than 17 percent in the crisis year 2009, hybrid sales actually went up. Around 350,000 were sold that year, pushing their market share up to 12 percent – compared to 3.4 percent in 2008. The third Prius, released in Japan in the summer of 2009, became the best-selling car of the year. The reason for this exceptional jump in sales was a Japanese government premium for environmental-friendly cars. Introduced in the spring of 2009, the scheme subsidized the purchase of a hybrid vehicle to the tune of JP¥250,000 (US$2700/€1900) through a package of tax breaks and direct subsidies. In addition to this, environmental consciousness in Japan is growing and, accordingly, the government extended the premium through September 30, 2010. In the USA, the design of the Prius, by itself, came to represent "hybrid" in the eyes of many. With its 50 mpg fuel economy rating, Prius is the most fuel-efficient passenger car available in North America, and in April 2011, the Prius passed the million mark in US sales.

Due to the high importance accorded to environmental protection and reduced fuel consumption in Japan, there is almost no domestic carmaker that is not investing in environmental-friendly powertrain systems, and so it is no surprise that Toyota's main rival in hybrid vehicles is another Japanese manufacturer, Honda. Toyota is under constant pressure from this smaller

Japanese competitor and its hybrid-lite models Civic and Insight. Their hybrid technology is less complex – the Insight only has a 10 kw electric motor – and the sales price is therefore lower. Furthermore, the development cycles and release of new versions are far more rapid than with Toyota's Prius, allowing Honda to regularly take a larger share of the hybrid market, even in the USA. As early as 1999, Honda was selling the two-seater mild hybrid Insight. The Prius did not hit America until 2000, and until the introduction of the second generation, Honda actually had a higher hybrid market share in the USA than Toyota. It took the Prius II to chase Honda's Civic from the top spot.[29]

Not only did the Prius II sell well, it had the thoroughly intentional side-effect of giving Toyota a shiny new green polish with no extra marketing at all. The vehicle was an international sensation, and made waves even in countries with low Prius sales, such the United Kingdom: "The car's surprise success has sparked a revolution in the car industry that is about to change forever the way the world's automotive sector operates," reported the BBC upon the car's release in 2004.[30] Prominent Hollywood stars such as Leonardo DiCaprio, Harrison Ford, and Billy Crystal drove the Prius to public events as a way of demonstrating their commitment to ecological causes, and this served to turn the Prius from product to cult in the USA. Prominent drivers of hybrids helped boost public awareness of the automobile's role in the environment, specifically putting a focus on improving fuel efficiency and reducing carbon emissions.

The trend was further strengthened by government policy during the Bush years, during which climate change was pushed to the bottom of Washington's agenda, making the hybrid into a very visible demonstration of alternative political views: while the Bush administration officially denied that climate change was either dangerous or worth combating through reductions in CO_2 emissions, the Prius offered ecologically conscious drivers the ability to set themselves apart from this stance. That was exactly the kind of self-generating effect that Toyota was hoping for. With its comparative hesitance in giving its products an individual brand personality (with the notable exception of Lexus), the green label applied to Toyota once mass-produced hybrids became widely available finally gave the Japanese giant a recognizable face.

What Toyota has learned, however, is that the powerful weapon that is the celebrity endorsement can also be turned against it. In the spring of 2010, for example, Apple founder Steve Wozniak complained publicly about braking issues in his Prius that, he believed, could not be just mechanical and almost certainly had something to do with electronics. This kind of comment from big-name drivers damaged Toyota just as much as ringing endorsements had earlier helped it to build up its green image.

Hybrid technology is nevertheless seen at Toyota as a core area of strength, and as a remedy – the technology that will help the company back out of the crisis. Whilst the recalls crisis and the earthquake have temporarily distracted the manufacturer from research, with Toyota dropping from second to eighth place in innovation rankings produced by the Center for Automotive Management, the manufacturer has been consistently placed in the top three in this list and it is inconceivable for Toyota to interrupt its hybrid research programs for longer than absolutely necessary. Toyota values the technology as it sees in it scope to combine the electric motor with practically everything – electric with diesel or petrol, electric with a combustion engine running on other fuels (e.g. biomass), or electric with hydrogen fuel cells – and Toyota is uncompromising in its aim to hybridize its entire range. Right in the middle of the sales crisis, in the summer of 2009, Toyota announced that it would begin manufacturing the Auris (the Corolla in European markets) as the first hybrid in Europe.

In January 2011, Honda pioneered a hybrid subcompact, which many observers saw as an experiment of sorts. The Honda Jazz was the first hybrid model in the 'supermini' B segment. Honda admits, however, that it is not as green as its rivals. Even with a new, continuously variable transmission, hybrid powertrain, and start–stop system, the Jazz emits CO_2 of 104 g/km, well above the rival Auris (89 g/km) and higher even than regular gasoline models such as the Ford Fiesta (99 g/km). Honda admitted that the Jazz could have been trimmed down to under 100g/km, which is the waiver threshold for the London congestion charge – but there would have had to be compromises along the way. Honda said this would probably have necessitated a larger battery, therefore reducing space in the cargo area; or the Jazz would have had to lose weight by leaving out sound-deadening materials, which would lead to a loss in refinement.

Of course, Toyota had to follow suit – and it did. The market leader announced that it would hybridize the Yaris, a competitor of the Jazz. Whether this leads to more satisfactory CO_2 results, with the Toyota hybrid system being much more complex and bringing in more weight than the Honda hybrid system, remains to be seen. Usually, experts believe that it is more effective to hybridize middle-sized or large cars, because there is simply more potential to save on fuel than with already economic small cars. At any rate, if Toyota succeeds in hybridizing its entire lineup within this decade, then costs for hybrids of all kinds may well sink even further. As many as ten new hybrids are in Toyota's pipeline and should be launched in the coming four years. Between the production of the first- and third-generation Prius, Toyota succeeded in cutting production costs in half, and the manufacturer has already set itself the ambitious goal of being able to produce hybrids at the same costs as diesel vehicles by 2015.

The hybrid strategy is nevertheless not without risk, since the vehicles are clearly not suited equally well for every market and every taste. Currently, Toyota's best results with the technology have come from countries where the state offers incentives to help customers purchase the comparatively expensive technology, as is the case with tax breaks in Japan and several US states. In markets without these sorts of measure, Toyota has had difficulty selling either core brand or Lexus hybrids. In the European Union countries for instance, the gasoline and related tax savings for customers are just not big enough to encourage the Germans to purchase the Prius, for example.

Furthermore, a hybrid vehicle can only really play its strong suit in urban environments, where electric driving is possible, and the car's fuel consumption and CO_2 emissions drop rapidly (of which the London study mentioned above is only the most recent in a long line of proofs). For long and fast highway cruising, however, even a Prius of the third generation is likely to be no more efficient than the average diesel; in fact, the weight of the two-engine car can mean that it uses a comparatively high amount of gasoline. This is probably the main reason why the Japanese and American successes with the Prius have not been replicated in Europe's largest car market: with thousands of miles of speed limit-free autobahns, Germany is diesel country after all.[31]

Toyota is getting ready for the next stage, however, with the Prius IV having broken through to the next stage of hybrid evolution: plug-in hybridization. This form of propulsion is proudly referred to by the automaker as "the best of two worlds." Compared to conventional hybrids, plug-in hybrid vehicles (PHVs) offer the advantage of being able to replenish energy reserves through domestic plug sockets. In the Prius, energy is stored by first-generation lithium-ion batteries produced in a joint venture with the Panasonic EV Energy Company (in which Toyota holds a 60 percent stake). Toyota anticipated various new technologies while the new Prius was in development, and this allows the prototype that has been under testing since the end of 2009 either to run on current hybrid technology (i.e. a combustion engine, an electric motor, and nickel-metal hydride batteries) or to be equipped with the new plug-in technology with the higher-performance and far more compact lithium-ion batteries.

The batteries Toyota is fitting in its Prius PHVs have a charge capacity of 5.2 kilowatt hours, which is more than four times the power current Prius models can store. Together with control electronics, the batteries weigh 140 kg (308.65 lbs.) and in order to fit them at the back of the vehicle, the floor of the trunk had to be raised by 3 cm (1.18 inches). The electric engine is designed to take the vehicle to maximum speeds of 100 kph (62.14 mph) and has a nominal range of 20 km (12.43 miles) when fully charged; after this, the gasoline engine takes over. Several hundred test models have already been given to customers all over the world, who are testing the performance, reliability, and durability of the new battery cells as well as general handling and the compatibility of the technology with everyday life.

As ever, Toyota takes things one step at a time, thoroughly testing and only releasing its products when it is sure that they are ready and it has gained enough experience of its own with them. If anything, Toyota is becoming even harder to pressure into action than before: in the wake of the recalls debacle, it knows that any serious technical issues with the new plug-in Prius would cause the giant another embarrassing loss of face. The model is not expected at dealerships anytime before the end of 2012, and by then, Toyota will also be offering solar carports for customers wanting to keep their CO_2 emissions as close to zero as possible.

The batteries that give the Prius PHV its 20 gas-free kilometers, however, will raise the price by a good US$6500 (€4500) compared to a third-generation standard Prius, and Toyota knows that even ecologically conscious customers will find this a bitter pill to swallow. On the other hand, in Japan and at least some US states, the carmaker will be able to hope for heavy tax rebates to get sales figures moving.

Without tax incentives, it is quite possible that, at first, the Prius PHV will not make much financial sense to consumers, even though it offers the possibility of electric city driving. Then again, the new Prius still has the potential to repeat the success of its predecessors, which are now part of the general accoutrements of environmentally conscious living. Toyota knows this, and has equipped the dashboard of the new PHV model with a monitor that can display tree silhouettes, "visualizing the customer's contribution to the environment" as Toyota explains it. Whenever the car is traveling on electricity, the virtual forest grows, and it is precisely this kind of gimmick that will loosen the purse-strings of particularly environmental-friendly customers when it comes to having to fork out extra cash for the batteries.

In any case, eco-conscious car buyers will find little alternative to Toyota any time soon. Daimler is planning to offer its next-generation S Class with a PHV variant, and the Chevrolet Volt or Opel Ampera, respectively, also feature a plug-in system with a combustion engine as a range extender for longer drives. General Motors, however, installed battery cells that give a purely electric radius of almost 60 km (37.28 miles), and these cost almost US$20,000 (€13.900); that's why the Volt/Ampera hybrid will be correspondingly expensive or, alternatively, a loss-maker for GM.

Toyota is counting on the same path to success in its PHV strategy as it has followed with all hybrids to date: it will throw the new models into the market at an unprofitably low price and rapidly increase production to high volumes. This has certainly worked for the Prius over the last three generations. After this Prius PHV, the next stage will follow, with the aim being to offer all Toyota models as hybrids by 2020. Toyota is counting on the double-engine car becoming its USP; it is supposed to haul Toyota out of its image and identity crisis and reinvigorate the automaker for the fight against the competition.

Electric cars: Japan concerned about staying on top

Developing the kind of high-performance, mass-producible, battery-powered vehicles required for fully-electrified road traffic is seen in Japan as one of society's overall aims. Just like China, the USA, and Europe, the island nation wants to be the leading market for electric vehicles, not least because it is a highly urbanized society with a large number of densely populated, built-up areas: the noiseless and possibly emission-free electric car would be perfect in this environment – even if its range is likely to remain low for quite some time.

Thanks to its years of experience with hybrid technology and extremely modern battery-production industry, Japan sees itself as being ahead in the worldwide race toward electro-mobility – still. Next to the Land of the Rising Sun, neighbors China, and, recently, also Korea are starting to become competitive in this field, and Japan is worried about whether and for how long it can hold on to the top spot. In order to respond to these concerns, METI founded a strategy commission in November 2009 in which experts drawn from politics, economics, and the sciences are discussing the "cars of the next generation." The commission has outlined a roadmap for further technological developments and METI announced that Japan would be aiming to establish its technology as the industry standard for electric cars.

Japan sees the development of very-high-performance batteries as one of the many keys to remaining a competitive industrial nation in years to come; after all, with its resource-poor landmass, Japan has long been well aware of the danger of overdependence on fossil fuels, and the search for alternatives is only going to gain in urgency. The island nation has long been very dependent on oil, and this is costly for domestic industry. The development of batteries has therefore been a consistent matter of national importance since the beginning of the twentieth century, and several members of the Toyota founding family encouraged it. This makes the fact all the more astounding that at one point Toyota, the industry leader in alternative powertrain systems, fell so far behind in electric car development. In a country investing so much in technology for mass-producible electric cars on so many levels, Toyota of all companies ignored this area far too long.

One of the most important preconditions for successful, widespread electro-mobility is suitable infrastructure for charging the vehicles, and in order to gather information about possible problems in this area when it comes to introducing electric cars on a wide scale, METI has been planning since

March 2009 to roll out 32,000 electric and plug-in hybrid vehicles along with a network of 5000 charging stations across a range of cities and towns. In March 2010, the Chademo initiative was founded, a federation composed not just of car manufacturers and energy producers, but also of charging station producers and electricity providers: their first aim was to produce a standard for a battery-charging station that could be adopted as a worldwide standard. As is usual in Japan, politicians, scientists, and industry representatives have come together to define clear goals in this area, and almost every company that could conceivably be involved in or affected by the standard has been consulted. The 158 members of Chademo read like a *Who's Who* of Japanese industry, with Nissan, Mitsubishi, Fuji-Heavy, and Japan's biggest energy provider Tepco all involved. The name is an abbreviation of CHArge DE MOve, a slightly mangled version of the English words "charge for moving"; in Japanese, it puns with *O cha demo ikaga desuka,* or "Why don't we have some tea (while charging)?" The initiative is tasked with installing battery-charging stations and establishing Japanese technology as standard; the idea is to make advances here so rapidly that the rest of the world is left without any choice but to adopt the system. The aim is that Chademo stations will be able to recharge car batteries so quickly that the most the driver will be able do in the time it takes is to drink a cup of tea.[32] Despite Toyota having fallen behind the competition for e-cars, it ended up participating in the initiative – mainly because stations will extend the electric range of its plug-in hybrids, too.

The Japanese government financed Chademo to the tune of JP¥12.4 billion (US$133 million, €92 million) in the 2010 fiscal year, but at Toyota many see possible difficulties with the idea of establishing international standards under Japanese leadership: "It will be a big and difficult challenge for the entire world to reach the same method in charging EVs. In the end, we may just need to adhere to the methods in each country."[33] As of late 2010, the number of charging stations in Japan is estimated at around 300, most of them in the Tokyo metropolitan region near town halls, shopping centers, and hospitals. The price of a charging station has been set at JP¥3.5 million (US$37,600 or €26,100), with the government shouldering half of the costs. There are three models available, with the fastest able to charge a Mitsubishi i-Miev battery to 80 percent of its capacity within 30 minutes. The smaller Subaru Stella battery is back to 80 percent after just 15 minutes.

An electric car made by Toyota

Toyota, too, has made consistent, tireless efforts to build an electric car that reaches its standards. Since 1996, the manufacturer has been experimenting – just like its competitors – with a variety of concepts, a development that was not driven by Japan alone, but also by the USA: several state authorities tried to influence car companies by offering various forms of support, such as electric infrastructure or tax rebates for environmentally friendly cars (see Chapter 1).

Moreover, Toyota's own standards and declared aim of building zero-emissions cars will force it to continue trying to build a fully electric vehicle. As part of the Global Vision 2010 strategy, Toyota coined the phrase "zeronize and maximize," with "zeronize" referring to things Toyota wanted to push down to zero – accidents/serious injuries, congestion, and indeed, emissions. The only thing the manufacturer wants to maximize, meanwhile, is the comfort and driving quality of its vehicles. Between 1998 and 2003, for example, Toyota offered a battery version of the RAV4, and more than of 1000 of them were built, with around 800 of them still in circulation today. These vehicles, however, were only saved from being scrapped by Toyota after a public campaign under the heading "Don't crush!" Toyota ceased production of the battery-powered RAV4 in 2003. In 2009, however, Toyota appeared to have come back to electric vehicles: it presented the iQ microcar study at the Detroit Motor Show. This electric vehicle is intended to be one of the pillars of future electric development, and its series version should offer a range of around 90 km (55.92 miles) on a full battery. It is a model aimed at commuters above all, and the cells will be produced by the Sanyo joint venture with Panasonic. Furthermore, by 2012, Toyota wants to have a new version of the electric RAV4 on the market, equipped with Tesla batteries.

Sanyo is one of the largest producers of lithium-ion battery cells in the world. Taken over by Panasonic in 2009, the firm has chosen not to specialize on any one buyer, but has nonetheless invested JP¥13 billion (US$140 million) in a new plant for hybrid-vehicle lithium-ion batteries in Kakogowa (in the Hyogo prefecture near Kobe, Japan). In 2010, the factory started production of over 1 million battery cells per month. In contrast to Sanyo, meanwhile, Panasonic is connected to Toyota via Panasonic EV Energy, a company that supplies Toyota with nickel-metal hydride batteries for hybrid cars.

To date, however, Toyota has been quick to point out problems in terms of the driving range and high battery cost of purely battery-powered vehicles. It is not reckoning with significant demand for electric cars until the second half of this decade, and this explains why it put the brakes on earlier electric developments in order to prioritize hybridization as a mass-market technology; and poor sales for purely electric cars such as the Volt, i-Miev and Leaf in 2010 and 2011 would seem to lend support to this view. Nevertheless, the competition is offering full electric, and Toyota knows that it will have to step up its efforts here, even if it sees true commercial prospects for the vehicles more in terms of decades than years. The strategic cooperation with Tesla is intended to mark Toyota's re-entry into electro-mobility.

Despite GM's high-profile entry into the EV market with the Volt, Toyota's strongest competitors in the electric cars arena are above all fellow Japanese manufacturers such as Mitsubishi and Nissan. Neither of those companies have market-ready hybrids, yet, but both are pushing into state-subsidized electro-mobility. The Mitsubishi i-Miev has been sold in Japan since 2009, costing around US$48,000 (€33,000); a third of the high purchase price is instantly refunded by the state, meaning that the end price for this electric vehicle for the Japanese car buyer is nearer to US$32,000 (€22,000). It is advertised as having a range of 150 km (93.21 miles) in energy-saving mode and 100 km (62.14 miles) with climate control, which should be more than sufficient for most Japanese commuters. The slightly longer US version of the e-car will be available in the USA by late 2011.

Nissan has entered a somewhat lower price category with its Leaf, a compact electric car with five seats and a range of 160 km (99.42 miles) that has been sold in the USA since late 2010. The advertised price there is US$32,780 (€22.800), but once tax rebates have been taken into account, the end-price should be closer to US$25,280 (€17.600), pushing down the local price of the Leaf almost into the same band as the Toyota Prius and a little higher than the Honda Insight.

Electric pioneers like Renault-Nissan and Mitsubishi are receiving government support in several other countries, too, especially those that have incorporated environmental incentives into economic stimulus packages in the wake of the economic crisis. France, for example, is subsidizing the purchase of an electric vehicle with €5000 (US$7200); Great Britain and China are following suit. It is not just due to this carrot that European car makers are trying to add zero-emissions cars to their ranges, but also to try to avoid the stick of EU-wide CO_2 limits calculated across manufacturers' fleets. Since it is unlikely that the EU will include emissions caused by power generation in the final balance (even though electric power is unlikely to have been generated without any CO_2 having been emitted, say from coal-fired or gas-fired power stations), electric vehicles counting as zero-emission vehicles are the perfect way for manufacturers to bring their fleet averages down.

The stake in Tesla taken in May 2010 opens up new paths for the automaker, as Tesla is currently the only company with a highway-capable electric car, but it had only managed to sell 1200 as of late 2010. In order to mass-produce the vehicle, the company needs a big partner. With the TPS and its admirable knowledge of efficient production, Toyota is well placed to help Tesla. In return, Toyota will expect to profit from Tesla's experience, both with purely electric drivetrains and their components, and with sales, service, maintenance, and repairs to electric vehicles. By 2012, the two companies want to be in a position to start mass-producing electric cars together, including Tesla's new family vehicle, the first premium electric sedan "Model S".

This alliance, forged between two very different partners, shows how much the industry has changed, allowing newcomers such as Tesla or the much bigger battery producer BYD (China) to attack the established car manufacturers on this new and still young field of individual transport. Toyota, however, has turned this threat into an opportunity and can now begin closing the strategic gap between it and some of its competitors in electro-mobility. However, even though purely electric cars have become an important pillar of Toyota's strategy for ecologically sustainable mobility, the central focus will nevertheless continue to rest on the worldwide success of the Prius and hybrid technology for some time yet. On this subject, the automotive giant never tires of stressing that, for many years to come, a combination of electric and combustion engines will remain viable for mass mobility; it may be the only viable option well into the future.

Fuel cells: the new hope for coming decades

On the way to zero-emissions vehicles, Toyota has already started trying to think around the topic of purely electric cars, which are in fact not totally emissions-free unless the energy they use has been sustainably generated. In order to ensure that any zero-emissions car really does not lead to CO_2 being released into the atmosphere, Toyota is counting on technology rejected by many other car manufacturers as impracticable – the fuel cell. Toyota finds it attractive, however, not least of all because it can be mounted next to an electric engine in a hybrid.

This alternative takes hydrogen as the source of energy, which is then released by the fuel cell, itself composed of a stack of galvanic cells that essentially do the opposite of electrolysis by fusing hydrogen and oxygen to make water; this, in turn, produces electric energy, which can then be directed either straight to an engine or to a battery. It is an extremely clean way of producing energy and has led to several car manufacturers producing concepts for fuel-cell vehicles over the last 20 years.

It would have been out of character for Toyota, with its emphasis on clean technology and future mobility, not to investigate this promising technology, and since 1992 it has produced six generations of hydrogen fuel-cell vehicles under the initials FCHV – fuel cell hybrid vehicle – each new generation better than the last and solving another few of the many problems so far presented by this technology.[34] The current generation of the Toyota FCHV was approved as roadworthy by the Japanese Ministry of Land, Infrastructure and Transport on June 3, 2008, and has been in (very limited) circulation since then. The concept is hybrid, with the vehicles carrying both the fuel cell and an electric motor. It has a tank storage capacity of 156 liters (41.21 gallons) of hydrogen, and at a constant tank pressure of 700 bar, the FCHV can cover 830 km (515.74 miles) on this amount. To date, only a very small number of these vehicles have been delivered to testers.

In 2010, Toyota announced plans to give more than 100 Toyota FCHVs to a US demonstration program over the following three years. The vehicles will be placed with universities, private companies, and government agencies in California and New York. Over the three-year course of the program, as more and more hydrogen stations come online, the intention is to add other regions. Toyota's fuel-cell program expansion is designed to provide one of the largest fleets of active fuel-cell vehicles in the USA with the primary goal of spurring essential hydrogen infrastructure development, prior to its 2015 market introduction.

In Europe, likewise, Toyota gave a small fleet of five to the German Federal Ministry of Transport in Berlin as part of the German–Japanese Clean Energy Partnership. In total, around 40 vehicles were actually tested in order for Toyota to gather experience using the fuel cells and the accompanying infrastructure (hydrogen stations, especially). The project was also financed by Daimler, Opel/Vauxhall, Ford, BMW, and Toyota's new rival, Volkswagen. The fact that in Europe, for example, so many manufacturers banded together for these tests indicates just how difficult hydrogen engines are; no single carmaker can claim to control enough of the technology to go it alone. As an energy source, hydrogen is complicated, expensive, and ecologically by no means uncontroversial: it may not create any carbon emissions when used as a fuel, but a large amount of energy is required to produce it in the first place. For more information on hydrogen, see the analysis in Section 5.4 in Chapter 5 of this book.

In light of some recent technical realizations about hydrogen's problematic nature, not to mention radically shrinking budgets, most automotive companies have either cut funds for researching this once so promising alternative propulsion technology, or simply cut it from their R&D programs altogether. Electric engines and batteries have been given priority due to the comparatively short timescale over which they can be developed into competitive products. This means that many now see a hydrogen fuel-cell car as a vision for the distant future, and even hydrogen optimists do not expect to see it in mass production much before 2050. As early as 2003, METI had published the ambitious goal of putting 5 million hydrogen vehicles on Japanese roads by 2020, and financed test projects to this end in which Toyota was usually one of the main developers concerned, but the difficulties proved insurmountable for the present. Toyota still believes in hydrogen, and thinks it could be another trump card, but is well aware of the fact that this will probably take several decades. In the meantime, "hybrids will bridge the gap between now and the arrival of hydrogen-powered fuel cell vehicles," says Graham Smith, Toyota's Head in Great Britain, explaining Toyota's strategy.

Even Toyota, with its legendary endurance, will not be able to jump the numerous hurdles on the road to hydrogen fuel cells in just a few years, and for this reason they are not mentioned in the strategy master-plan to which Toyoda frequently referred when presenting figures for 2010. Yet it would be unfair to suggest that Toyota has dropped hydrogen or that it will be incapable of ever making it work, and many competitors are becoming increasingly interested in the progress the automaker has made in this area; they are looking to cooperate in an attempt to manage the unusually high research costs. Daimler, for example, could well be a promising partner in this area, since it now has a Tesla stake in common with Toyota and makes no bones about being in the market for cooperative initiatives in hydrogen. Since 1994, Daimler has invested more than €1 billion in this technology and industry observers estimate Toyota's investment in it to be roughly the same. Both manufacturers have announced series production of fuel-cell vehicles for 2015, with Daimler aiming to adapt its B Class, and if they were to enter into cooperation, considerable synergy effects could well be utilized for the first generation.

Despite the billion-yen loss of 2009, the never-ending series of recalls, and the 2011 earthquake, Toyota has by no means renounced its goal of being the world's most environmental-friendly carmaker, and certainly has not abandoned its strategy for leading the field in terms of alternative energy cars, either. Quite the contrary: the potential that dwells within the world's biggest carmaker has certainly not been evicted by recent difficulties. Toyota will keep working on hydrogen – just as it will on all of its other alternative strategies – step by step, slowly and patiently, until the technology generation after next is being asked for and the company is once again revealed to be streets ahead of the competition, as has been the case with hybrid vehicles.

Toyota, the wounded giant, is in many ways more of a phoenix, having risen from the ashes of 2009 and its record losses back into profitability in 2010. With an operating profit for 2010 and sales in the USA recovering, the giant had got a spring back in its step after a long time spent reeling, stumbling, and limping. While the recalls had slowed the giant's recovery, it became clear that they have not stopped it altogether: in calendar-year sales figures for 2010, Toyota still managed to top its major rival, General Motors. Albeit only by a couple of thousand cars, Toyota remained number one carmaker in the world, having sold more than 8.4 million vehicles (including Daihatsu and Hino) in more than 180 countries, including China. Then the giant, just getting back into its stride, was once again sent reeling by the Great East Japan Earthquake, allowing its competitors – especially GM and Volkswagen – to steal a march on it. Now it is once again up to Toyota to prove that it can be something more special than a phoenix that rose from the ashes – a phoenix that rose from the ashes twice.

4

THE RACE FOR POLE POSITION: WINNING BY A NOSE

In the race for the top spot in the automotive world, Volkswagen and Toyota have pulled ahead and pushed their competitors in categories such as sales, revenue, or technological leadership back to the second or third ranks. Volkswagen is now making gigantic leaps in an effort to overtake Toyota – the reigning champion it had called out four years ago and which has suffered so much since. American readers, meanwhile, may wonder where GM – for at least 70 years the world's biggest carmaker – figures in this duel at the top.

4.1 PHOENIX FROM THE ASHES

In the darkest days of 2008 and 2009, almost no one could have foreseen GM's sudden and vigorous resurgence in the course of 2010. Yet, slimmed down and freed of crippling debt obligations by the US Government, GM was better placed to profit from the recovery in its home market than it might have otherwise been. Not only the Government intervention and improving economic conditions, but also the quality issues facing its rival Toyota helped GM (as well as Ford and Chrysler) to double-digit sales growth throughout 2010. GM even managed to return to the stockmarket at the end of that year, and its stock price has held up well. Nevertheless, even this kind of recovery can not make up for lost time, and GM's difficulties began long before the financial crisis, writing year after year of record losses until the bankruptcy. Despite consistently high sales figures (which were, however, mainly bought with incentives) GM has been financially weighed down by what are known as "legacy costs" – or retiree health and pensions programs. Although some of these liabilities have been settled in the course of the restructuring, generally speaking, it is still the case that for every 1000 members of staff currently employed GM has in way or another to finance pensions and healthcare for between 5000–10,000 retired workers and their family members. By some estimates, the legacy funds continue to be underfinanced to the tune of US$2 billion annually, and this kind of cash strain is a hefty burden in a time where research and development bills are skyrocketing to meet the demands of next-generation power trains. GM is trying to make up for lost time, but is being slowed down by the burden of the past.

To outsiders, of course, the sales figures look great, especially in the habitual comparison between GM and its Japanese nemesis, which lost market share in the USA in 2010 and – shaken by the Great East Japan Earthquake in March 2011 – has had trouble supplying the USA, its biggest market. Yet Toyota is not likely to remain weakened for long: announcing its 2011 half-year results, it said that production, "already at 90% in June", would return to "near pre-earthquake planned levels in July", with "full production able to flexibly meet consumer needs" by the close of the year. Further to this, the firm has made clear its intentions to buy back market share in the USA by every means necessary. This means one thing above all: the heavy use of discounting and incentives, once one of GM's core weapons. GM has every reason to be worried by this announcement. In its quest to keep its home factories working at full capacity, Toyota will push prices in the USA to new lows; but due to its lean set-up and full reserves, the Japanese company will be able to handle this kind of price-war. GM, however, cannot afford to hold out for long, and there is the ever-present danger that their network of dealers will be unable to stomach this new blow. Furthermore, there are signs that GM is falling back into bad habits, with its dealers currently sitting on a backlog of 120 days worth of production in expensive heavy vehicles. This contrasts with an average backlog of 80 days of sales supply in the same category at its competitor Ford and indicates a difference in market discipline between the two.

That GM (and Ford) has been able to hit back at Toyota during its difficulties at all is a testament to the success of the restructuring work carried out by American automotive executives. The bloodletting at the American Big Three has gone on for long and been so grave, however, that even the current infusion of cash – both from Government and from increased sales – still leaves them looking somewhat anaemic. Furthermore, it is not just a question of financial resources, but of time: "With the bankruptcy, we lost roughly a year in terms of development," said GM CEO Akerson in early 2011. Despite glossy press images of the new flagship models, the Volt and the Ampera, which are technologically advanced but neither pure EVs nor big sellers, GM is still faced with an exceptionally tough environment going into 2013–2014, during which time their rivals VW and Toyota are planning to release a slew of new models the likes of which have not been seen in automotive industry. Not only are their cars brand new, but there is a highly credible production strategy for the future behind them: this is something that GM, trying as it was to keep its head above water for so long, has had no opportunity to develop to the same degree. Furthermore, GM is as reliant as ever on SUVs and pick-ups – vehicles which may well be past their prime even in America as the era of cheap gasoline seems to be drawing to an end. In view of these difficulties, it seems clear that the only two automotives with potential to reach the top of the worldwide industry in the medium term are Toyota and Volkswagen.

4.2 ROUND TWO

It is not for the first time that these two unequal rivals have been so close. In 2000, Toyota and VW both sold around 5.2 million vehicles worldwide, finishing that year in third and fourth place respectively. However, in the following years the manufacturer from Germany just could not sell above the decisive mark of about 5 million vehicles. It took VW another five years and a cost-cutting program before they were able to start moving toward the 6 million mark from 2005 onward. Meanwhile, the Japanese were pressing on toward 9 million vehicles sold per year, earning them the top place in the industry by the end of 2008.

Nevertheless, the sales slump in 2008–9 and the huge recall crisis in 2010 forced Toyota to turn away from its relentless expansion policy. With Akio Toyoda's humiliating declaration before US Congress that Toyota had grown too fast, a new era in Toyota's history was beginning – and an old chapter was brutally closed. The industry leader has learned that size alone is no guarantee of success, that it can even turn out to be something of a curse – and in more ways than one. For one, once the boom turned to bust, Toyota's enormous production capacities turned into liabilities and tore multibillion-yen holes in its accounts; and the sheer size Toyota had grown to is widely seen to be the central reason for its failure to spot possible quality issues before they turned against the company and bruised its image dramatically.

Toyota had been proud of having overtaken the US manufacturer General Motors as the world's largest carmaker in 2008. Now, of course, Toyota sees things very differently: it is quality and leadership in green car technology that the Japanese are aiming for and, as Toyoda proclaimed, excellence in these fields is not a question of size. Sales of 7.5 million vehicles, plus the vehicles produced in joint ventures with the Chinese: this is the target to which Toyota has lowered its break-even mark in terms of profitability.

Meanwhile, Toyota's rivals from Detroit have risen from what appeared to be their deathbeds and are making Toyota's life difficult in the USA. Whether this stellar comeback of the Detroit manufacturers is sustainable, remains to be seen. At any rate, both Toyota and VW are facing fierce competition from rivals old and new: the Korean manufacturer Hyundai, for example, plans to sell more than 6.3 million cars in 2011, almost as many as the whole VW Group sold in 2009. What makes Hyundai particularly dangerous compared to manufacturers in the same volume category is their extraordinarily high level of both growth and profitability over the last five years.

However, the crisis has taught Toyota a lot, and perhaps the most important lesson is that no problem, big or small, can be ignored or left to deal with later once it has been recognized. Toyota seemed to have lacked clear and swift top-down leadership and experienced the damages of weakness and delayed action. Yet the recalls showed that the potential damage turned out more costly than an early (and perhaps expensive) intervention would have been. In the aftermath of the 2011 earthquake, this realization may have helped Toyota react as fast as it has; steeled against the challenges posed by the catastrophe, Toyota is on a mission to keep its place in the worldwide automotive industry whilst being the pillar of Japanese society it has always been. Drawing on the typical Japanese characteristics of discipline and patience, Toyota seems able to surmount any crisis, even one as devastating as the events of March 2011. The first priority now is to get its factories going again. The next goal will be to return to previous profitability levels and from there, to the top spot: and it only seems a matter of time until the company manages it. If Toyoda remains CEO, he will now have a vast wealth of experience in crisis management and will continue to implement the much leaner and more decision-oriented management style he has become known for. Pre-earthquake announcements about capacity reductions in the homeland – estimated in the region of 700,000 units in Japan by 2017 – would then return to the agenda. The CEO drastically shrank the number of board directors from 27 to 11 – the first ever such cut at Toyota – and gave more autonomy to the heads of the regional organizations: a clear lesson from the recall crisis.

This stands in stark contrast to the mood of the German competitor. Spurred on by the tie-up with Porsche and fired up by the new possibilities opened by the modular toolkit approach, Volkswagen is going all-out for growth. The German group sees size as a crucial pillar in its vision for the future, and CEO Martin Winterkorn is not concerned with disguising this: the German contender has set itself a target of selling 10 million vehicles annually by 2018 at the latest, and it seems (after a record fiscal 2010) this target will be reached much sooner. The plan calls for 1 million vehicles yearly to be sold in North America, at least 3 million in China, half a million both in the ASEAN countries and in India, and a further 5 million spread across South America and the almost completely saturated, low-growth European market. If Volkswagen does want to grow organically, as Toyota did through the early years of this century, it will have to average 360,000 extra vehicle sales every year through to 2018.

The year 2011 is very likely to have brought Volkswagen considerably closer to this goal. While Toyota has defined a new bottom line for profitability in the midst of the re-call problems and earthquake disaster, VW is betting on world-wide economic growth as one of the main drivers to boost its business.

4.3 VOLKSWAGEN MOVES TO OVERTAKE

Volkswagen now is talking the talk in terms of its corporate goals for 2018, and for this Group, full of conflicting interests, management knows that talking the talk – especially to its own staff – is a necessary prerequisite. The kind of quiet, tip-toed, non-confrontational style that Toyota consistently strives for would lead to sheer anarchy in a Volkswagen Group eventually to be composed of ten different brands, not to mention other interest groups such as the influential union IG Metall and the State of Lower Saxony. Nevertheless, the rough-and-ready rhetorical style practiced at Wolfsburg, as necessary as it is to encourage the employees, will not by itself achieve the 2018 objectives. A *de facto* evaluation of the 2018 goals will have to go beyond rhetoric and weigh up the strengths and weaknesses of the Germans in the face of manifold upcoming challenges in the car industry. Beyond that, VW begins to see the dangers of arrogance: complacency, encouraged by record sales figures in 2010 and 2011, could be one of the greatest threats to Volkswagen's ambitious goals.

Volkswagen has to compete against the leader in hybrid technology, and the ecological objective was added to existing plans only in a final step: until recently, the high costs for green car technology and the market-leading progress it had made in developing efficient gasoline and diesel engines had lulled Volkswagen into a false sense of security as regards hybrids and electric cars. Now, the Group is very publicly making the running in a field in which for so long it placed no belief. After the failure of the Audi Duo hybrid in the late 1980s, VW slowed any development of hybrids; projects in this area were mothballed or saw their funding cut and Wolfsburg often publicly proclaimed that hybrid technology was only suited to a megalopolis with average traffic speeds of around 10 km/h (6.21 mph) as well as stop and go mode. This attitude meant that only a very limited number of hybrids ever even made it beyond the concept stage at Wolfsburg, and VW managers did not tire of mentioning the disadvantages of gas-electric cars, particularly in view of German driving preferences. As a consequence, Volkswagen has precious little experience with the application of two different engines in its cars, and this limits its capacity for creative and quick problem solving once difficulties with technical details in coming generations have to be overcome.

In fact, Volkswagen's first hybrid series only arrived on the market in 2010, and hybridization will mostly be featured in its high-end models such as the

Touareg SUV where the price differential caused by the hybrid powertrain system is not as noticeable. VW also announced the launch of a hybrid version of the Jetta, which is intended to challenge the Toyota Prius, which by then most probably will already be on sale in a plug-in version. Whether VW can really put the bite on Toyota's by-then fourth-generation hybrid system remains to be seen.

Volkswagen has not been an exception in German petrol-head culture, however. In general German manufacturers spent far too much time and energy trying to stop European Union regulation of CO_2 emissions, which was viewed as a plot by the neighboring French high-volume manufacturers to neutralize the German strength in the premium classes (which it probably was). In this climate, investments in hybrid powertrain systems were also unpopular with suppliers, who nevertheless will play a crucial role in the German car industry's jump into the hybrid and electric era. Nowadays German manufacturers are forced to catch up quickly, and their suppliers, at first shaken and now strengthened by the crisis, will not let them have alternative technologies cheaply. Costs for any future innovations will most likely have to be shouldered by the carmakers, too.

Toyota, on the other hand, not only has a very strong hybrid brand identity, but is already at the next level of technological evolution with the PHV. Driven by the desire to spread sustainable motoring not just across its home islands, but the world as a whole, Toyota is sure to pin the environmental imperative to their flag, making their recovery not only a matter of national pride, but also a text-book comeback-with-a-moral story. The deep pockets and astonishing resoluteness of Japanese society will provide the environment in which new technologies like plug-in hybrids can really flourish – as early as 2012, Toyota will be fielding a vehicle that can be recharged via domestic plug sockets. Just how well it will sell depends on two factors: how fast the price of oil rises and which governments will then be subsidizing the purchase of a PHV.

Whatever the sales figures, fuel consumption will continue to be one of the major measures by which Toyota defines the success of its next-generation cars. It has managed to save roughly half a liter of fuel per 100 kilometers between the second- and third-generations of the Prius, and its fourth-generation vehicle is planned to achieve average consumption of just 2.6 liters per 100 km (90.47 miles per gallon). These are figures that Volkswagen simply cannot match on the volume level in the short term, even with its newest models in the compact segment. The VW Blue Motion Polo may perform at respectable 3.3 liters (71.28 mpg), on a level with the Daimler Smart CDI, and VW is demonstrating its eagerness to catch up with concept cars such as the XL1. Meanwhile, the diesel plug-in hybrid vehicle represents the third evolutionary stage of Volkswagen's 1-liter car strategy and features combined fuel consumption of 0.9 l/100 km (261 mpg US), according to VW. Still, it is only a concept, while Toyota will soon have its plug-in model on the streets.

The Japanese simply have far more experience in this field, and there is little to stop them from keeping ahead of their competitors. In contrast to VW, currently taking its first steps with hybrids, Toyota is already profiting from its high-performance, low-price drivetrains as well as good link-ups with battery producers, who will supply them with the all-important lithium-ion cells in the foreseeable future. Smart battery technology and high-performance power electronics: these are the areas in which Toyota is ahead of its competitors and produces more cheaply, exploiting long-term economies of scale and continuously pushing down costs for hybrid powertrain technology. None of this came cheap – Toyota is reported to have sunk US$10 billion into R&D for its hybrid system – and only Toyota itself knows whether this money has already been earned back by increasing hybrid sales. Volkswagen is still stuck with a relatively expensive hybrid system and would have a long way to go in order to match Toyota's healthy position in the green market.

The top managers at Wolfsburg, however, plan to capitalize on high-volume modular toolkits in order to reduce the costs of their hybrid technology: Winterkorn wants to "pull hybrid out of its niche" and is pushing his engineers to develop modular toolkit approaches that can be applied not just to one solitary hybrid flagship, but to several models in the range. In other words, numbers will be the key. This will not come inexpensively, however, either in investment terms or with regard to risks for the VW brand image: it will take a lot of effort to maintain the characteristic riding qualities of a Volkswagen, and the engineers are well aware of the fact that this is something that the Group's customers value. Nevertheless, Winterkorn's words will force VW to accept eco-car strategies, and management will have to go along with it. Another important step will involve suppliers, because VW relies on their own capacity for innovation and the range of hybrid components they can offer to a far greater extent than Toyota.

For VW in the USA, the first tangible result of the change will be the Jetta, developed especially for the US market; the hybrid version has already been dubbed the "Prius-killer" by automotive journalists. According to industry analysts it will be hybridized by adding a 20 kw electric motor. Yet, comparisons of this to the new Prius show just how far behind VW is at this stage: Toyota is already offering 60 kw – i.e. triple the power – and in order to make sure that the car still feels like a VW, the German group is forced to run it on an expensive mechanical gearbox rather than the hybridized transmission installed in the Prius. Nevertheless, this does put VW in the same category as Toyota's nearest rival, Honda, whose flagship hybrid is the Insight, carrying a similar technology. This shows just how much Volkswagen will have to catch up before it can bring pressure to bear on its declared rival. Not that the Group is not trying to bridge this gap as fast as possible: according to the Centre for Automotive Management (near Cologne, Germany), in 2010–2011 Volkswagen booked by far the highest number of technical innovations of any automotive company, with 128 novelties, 50% of which had to do with engines. In this field, it was one of the only companies worldwide to produce more innovations relating to alternative (and especially hybrid) technology than for conventional engines.

4.4 TOYOTA: DIGGING IN ON THE GREEN FRONT

Toyota is quietly arming itself against its challenger. Much like Volkswagen with its modular system, Toyota has a strategy to build on economies of scale spread across the entire lineup: eventually, the Japanese will hybridize their entire range, offering a hybrid version of each model. Detailing its new eco-car strategy, the pioneer announced that it planned to launch ten new hybrids by 2015. Meanwhile, Toyota has turned the flagship Prius into a hybrid model family, offering model variants for families or single, urban customers, in an effort to make the hybrid system attractive to a broader base of customers. If the hybrid strategy succeeds, this can only bring further cost reductions for the Japanese carmaker and an even higher share of the hybrid market in the USA and Japan, securing the manufacturer's top spot, while keeping competitors at bay.

Meanwhile, battle lines for the purely electric mass market car are being drawn up differently. New challengers – such as the Franco-Japanese Renault-Nissan alliance and that of the Japanese manufacturer Mitsubishi with Peugeot/PSA – are pushing past the two biggest carmakers and, while both Volkswagen and Toyota have produced purely electric vehicles as concepts for a long time, neither had taken the next step and set electro-mobility as a major, immediate priority with a corresponding distribution of resources. Regardless, however, both carmakers are pushing ahead with ambitious plans for e-mobility.

In Wolfsburg, the new e-car strategy is crucially based on sales prospects in the Chinese market which are slowly crystallizing around battery-driven mobility thanks to massive political pressure from the government. In the People's Republic, state purchase premiums are intended to bring the price of battery-driven vehicles down to the level of conventional cars – the huge country hardly has another option: if this nation of 1.3 billion inhabitants allows too high a degree of gasoline-based mobility, the environmental effects would be absolutely devastating. And this should put no small amount of impetus into a German-Chinese partnership that has already proved beneficial for both sides and will do so again if electric cars become a realistic alternative for the Chinese megacities. Both joint ventures in which Volkswagen is engaged are planning to sell their first mass-market electric vehicles in 2013 or early 2014.

At a global level, however, electric vehicles will play a relatively unimportant role in the immediate future. Outside of China, Japan, and a few particularly environmentally friendly US states, battery-powered passenger cars will be a niche product until at least 2020. In their current form, their range is simply too limited and their batteries are too expensive for most customers and markets.

Toyota is still betting on its hybrid card, but at the same time is pushing ahead with its new partner Tesla and new all-electric models (the cooperation with Tesla allows Toyota to experiment with the new technology without risking its own already damaged image). This strategy is aiming at the American market, among others, where a number of states (including California, Toyota's US headquarters) are fostering electric mobility using tax rebates or building charging facilities, for example. If Toyota continues with this project, it could be another major boon to the carmaker's image as a green manufacturer.

However, both VW and Toyota are likely to end up in the middle of the pack when it comes to electric vehicles rather than shooting ahead. Both seem to have accepted that they are fast followers rather than pioneers in the field. In view of the high costs and low global sales for this relatively new automotive product, this will be of little immediate concern to either, however.

In the medium term, Volkswagen's strategy for conforming to EU emissions legislation is to successively shrink engine displacement while increasing efficiency. VW is still a world leader in conventional diesel and gasoline engines, placing its hopes on high-pressure direct injection systems. The new two- and three-cylinder engines under development will allow the Group to maintain its lead in conventional technology; meanwhile, the German car maker still has enough time to catch up with green car technology such as gas-electric powertrains, or to hammer out alliances with cell manufacturers for electric cars – and will probably have the advantage of being behind the pioneers on this occasion. Moreover, VW can realistically hope that its optimized engines will sell in high volumes and continue to offer decent profits for several years to come. Certainly, profits from conventional models will still be higher than those from expensively developed green cars.

4.5 THE BATTLE ON THE BOTTOM LINE

After all, if Wolfsburg really is to overtake Toyota, it will also have to start solving some problems other than green car technology.

One of Toyota's major strengths remains its ability to produce mass-market cars at exceptionally competitive costs, and on a global scale, despite the fact that the company – like Volkswagen – has an expensive production home base. The VW Group, on the other hand, still produces a much higher proportion of its vehicles in-house than the core organization of Toyota with its *kairetsu* network. In view of the high degree of participative management at strategic level and the political aspect of having Lower Saxony on the Board, it will be obliged to continue producing all the way up and down the value chain in Germany, despite the country's comparatively high employment costs. This is a major disadvantage for the carmaker, leading to significant potential losses in profitability. Whichever figures are compared – *Harbour Reports*, profit margins, or the average vehicle output per employee – VW lags behind Toyota (with the exception of the earthquake 2011). In 2006, for example, Toyota produced 28 vehicles per employee; VW managed 17. Until the crisis, Toyota's productivity at home in Japan rose consistently; VW was not able to kick-start its productivity until 2006, which was mainly a result of the restructuring by the controversial former VW manager Wolfgang Bernhard. As a result of the Earthquake and the preceding crises of 2008–2010, Toyota will need a few years to come back to this level of productivity: but come back it most surely will.

Volkswagen may proclaim that it is now making progress, increasing productivity by at least 5 percent annually; but set against a benchmark lean manufacturer such as Toyota this is not enough to close the productivity gap. To close it, some serious taboos would have to be put under the spotlight in Wolfsburg, with sacred cows such as in-house components manufacturing being assessed purely objectively. This would be good for profits, but is improbable for political reasons.

With its disadvantages in terms of overall productivity, Volkswagen is by no means untypical of German manufacturers: there is no shortage of experts to point out that, even with rising sales and turnover figures, German carmakers have repeatedly been incapable of achieving profit margins that rival Japanese or Korean manufacturers. As early as his first *Report* in 1989, James Harbour wrote of his doubts about the German car industry's ability to cut costs resolutely enough to gain handsome margins. Yet this is where Volkswagen has surprised the industry, with the 2011 figures indicating that they indeed have been matching Asian standards. Driven by Audi's fantastic performance and unprecedented factory utilization in China, Volkswagen delivered not just on turnover, but on the profitability promise too; yet whether the Group can keep this up through the risk of the introduction of the modular toolkits remains to be seen. Another issue which could see Volkswagen's gains melt away again would be another large-scale recession, which would be particularly dangerous if it reached China.

Behind the differing production approaches and cost structures of Volkswagen and Toyota lay entirely different product philosophies: the lean mass-production system is competing with VW's premium product in a mass-market approach. Toyota's philosophy is based on a never-ending crusade against waste, complexity, expensive but not exactly punchy innovation, and frictional losses caused by organizational procedure – especially on the factory floor. The Japanese manufacturer checks new designs and parts early on to make sure they can be mounted in just a few simple, efficient steps. Expensive proliferations of model varieties and versions that increase costs for every process from development through to tooling and parts stocking are nipped in the bud. The core of this philosophy is simple: products are adapted to production, not the other way around. This allows Toyota to offer generally dependable vehicles at bargain prices. Improvements to production processes are carried out with positively missionary zeal, keeping them stable and efficient across every link in the value chain. Toyota's dream is to design the car and the factory in union, enabling the carmaker to tailor every single step, every last movement to its ideal, bare, simple core, and reduce costs as a result. Toyota wastes no time adding amenities to its cars that might escape customers' eyes and therefore be hard to sell, particularly if they require a price premium. This strategy has turned out to be highly successful everywhere except in the European markets with their affinity for technological finesse.

Volkswagen's philosophy is in almost every respect the exact opposite. Here, passion is reserved for technical tinkering, perfectionism in the way the vehicle handles, and attractive car design. This is not just a passion, but a necessity, since every vehicle produced by the Germans, especially under the core VW brand, has to encourage customers to pay a premium. Volkswagen builds cars that make customers desire them, which in turn makes them dig deeper in their pockets, and technical refinements play a major role in this product philosophy. Nevertheless, this premium approach is not a matter of historical chance, but has been consciously established as a claim by the leading managers Piëch and Winterkorn through the Group's tie-ups with Audi and Porsche, all from southern Germany. Since the dawn of the industry, Germany's southern manufacturing regions have been synonymous worldwide with the premium class of passenger cars.

Carrying over these premium-class qualities to the mass market is what Volkswagen is about, and the disadvantages of this strategy become clear with a close look at profitability ratios. One-off successes such as the Tiguan – which turn in individual-model double-digit profit margins – are admirable compared to other models in the range, but the Volkswagen brand has in general remained remarkably unsuccessful in terms of profitability. This means that these occasional profit makers have little effect on the bottom line of the VW brand, which stands for more than half of the Group's sales. Even in the record year 2010, the core brand only reached a pre-tax return on sales of 2.7 percent. Compared to 7.1 percent for the Group as a whole, it becomes clear that the smaller brands – especially Audi and Skoda – are the ones that bring in decent profits, while the VW brand remains weak in terms of return on sales. Improvements in profitability would entail cuts in VW's home base, and this sharply conflicts with the Group's sacrosanct objective of securing jobs, which is even inscribed in the Group's statutes.

Just like most other German manufacturers, VW has to reconcile the conflicting objectives of improving productivity – in order to maintain economic competitiveness – on the one hand, with the political postulation that productivity increases should not involve major cuts in employment on the other. The powerful IG Metall union, for example, is unwilling to accept job losses. The unresolved conflict between profit and social responsibility is a major strain for VW, which is greatly influenced by political shareholders. Toyota faces similar political demands in Japan and until recently has solved the issue through aggressive growth – a strategy which VW is now following. With ever-increasing production and sales, Toyota never had a surplus of employees; moreover, staff could always be transferred to suppliers or temporary workers shed if circumstances demanded it, and there is no union opposition to speak of. Now, facing painful adjustments to the changing competitive environment, as well as managing the aftermath of the 2011 earthquake and the effects of the rising yen, Toyota has yet to prove whether it can reduce excess capacity in Japan. In the short term the economic situation in the motherland will prevent Toyota shedding too many jobs in Japan, but the strain this may place on the manufacturer should not be underestimated.

VW, by contrast, is now in its major phase of extreme growth, the kind of relentless expansion Toyota has just got over and done with. VW sees itself as perfectly positioned to use the momentum from returning global economic growth to overtake Toyota. Yet even if it does reach the sales volume targeted by management, VW will not automatically become the top car manufacturer: it will need to sort out its difficulties with productivity and profitability in the long term. It will be crucial, in this respect, to ensure support of the influential Works Council and IG Metall. In this context, it is interesting that it is the new US factory – with no union representation to speak of – that is supposed to be the most modern and efficient plant in the Volkswagen stable. From here, where they intend to set a productivity benchmark, the Group wants to "roll back" lean processes to its existing sites in Europe and elsewhere. The manufacturer plans to implement the "Volkswagen Way" as a match for the Toyota Way across its global empire. It remains to be seen whether VW will be able to match the Japanese for truly efficient production in Chattanooga; it is certainly hard to see the traditional central production site in Wolfsburg getting to that point.

The power of tradition is strong in VW boardrooms and shop floors alike, and the usual unpredictable perfectionism right before the start of production – a part of time-honored VW product philosophy – is becoming an increasingly heavy burden for the Group to bear. This custom of taking plenty of time to optimize a product again and again destabilizes the technically complicated launch of new models and can often cost significant money – more so as life-cycles of cars get shorter and product relaunches follow ever quicker. If the Group really does grow according to plan, this fanaticism for obliterating faults right up to the last minute could be enough to exhaust a company that is already putting considerable effort into revamping its model lineup and occupying niches that it has previously left to the competition.

4.6 THE TIDES OF GLOBALIZATION

Things will get even less controllable for VW because, in order to reach its target of 10 million sales, this multinational corporation needs to globalize itself even more, just as Toyota did after the two rivals parted company at the 5 million-vehicle mark. VW plans to generate the majority of extra sales in emerging markets – the BRIC countries, above all – and the USA, and these are the countries where VW is really ramping up its efforts. Yet, China aside, until recently the VW Group did not have a systematic approach like the one Toyota applied early on. The Japanese rival did not invent the turnkey factory just to help it advance into new markets with unusual speed; its "transplant" strategy also allowed it to expand in a standardized way, thereby eliminating errors in production processes. Toyota has standardized not only its products but also its production, and this is the key to its speedy growth and competitive costs.

Yet Volkswagen is not hopelessly lost. While Toyota concentrated on standardization of all processes from production through to management in order to increase its competitiveness, Volkswagen wants to use its modular system to set standards for components and continue to increase the variety in its model range while simultaneously reducing costs. This expensive development project could just turn out to be Volkswagen's secret weapon. It would allow for a high degree of standardization over its ten or more brands that would not be visible, however, to the brand-conscious customers, and for both development costs and time to be cut; and since the brand leaders in the Group have experience in using design and communication to differentiate their products and charge them with highly emotional values, they should be able to keep cannibalization between brands in the Group down to a minimum. If VW manages to implement this system according to plan it will gain a major strategic advantage against Toyota.

The modular toolkits would compensate for many of VW's current productivity issues while enabling the Group to order far higher numbers of (standardized) components from suppliers, with the positive side-effect of gaining pricing power in bargaining with those suppliers. If, however, defects were to creep into the new modular toolkit system, made up as it is of standardized functional units, they would spread like wildfire across the entire range of models and brands which all use the same components. This would leave VW open to the same risk as Toyota, whose standardized production meant it had to recall more than an entire year's production due to major problems with a gas pedal, among other things.

The modular system will be a critical success factor when Winterkorn and his team start to move into the last territories on the map in which there is, as yet, no Volkswagen flag flying. If it wants to swing itself into the leading position, the Group will have to continue to generate high turnover on the back of the global economic recovery – and to do so in all relevant car markets – just as Toyota did in the past decade.

In terms of market potential, however, growth accelerates the low-cost segment more than it does the premium ranges, and not just in the emerging economies. Yet VW has problems at the bottom end of the price scale and, to date, has been less successful than Toyota or Hyundai in offering the kind of good-value models the hotly contested emerging markets are thirsting after. The modular toolkit system may well allow for incremental graduation in individual models, tailoring them specifically to the requirements of each market; and Volkswagen is also developing a new "world car" that uses an existing platform in order to increase its profit margin. As the alliance with Suzuki has turned out rather complicated in everyday business terms, Volkswagen has to draw on its own resources in order to defend its market share against the aggressively growing Korean brands. Skoda boss Vahland wants to explicitly position his brand against Hyundai for example; in this plan, the new world car will play a strategic role. Produced under the Skoda brand, it is intended to be manufactured and sold globally in 2012. The new car will start as a five-seat sedan; later on an entire vehicle family is planned. The price of this Global Compact Car is expected to be no higher than US$17,300.

The Chinese joint ventures could manufacture further offshoots of the world car, some of which will be sold under the VW brand. On the other hand, the world car would be perfect to launch the new China brand, which the Chinese government has demanded and VW is mulling over. The fact that Skoda is leading the development of the new world car is at the same time evidence of the difficulties in the VW–Suzuki alliance. Another option Volkswagen has long had on the agenda is to cooperate with Suzuki in order to manufacture cheap, entry-level models for emerging countries such as

India. This comparatively small Japanese company is already manufacturing micro-cars across the world at very competitive costs. A modular toolkit for low-cost compacts and subcompacts was thought to be advantageous for both sides, but seems a long way off at the moment.

To date, Toyota has been far closer to the US$10,000–12,000 price category in the micro- and compact-car segment, with the Japanese having already managed to turn a profit building the Aygo in the Czech Republic and conquering the price category occupied by the Dacia brand of Renault. While Toyota still lacks good-value entry-level products for the Chinese market, it is already rolling out the Etios in India which is reported to be priced at INR525,000 (or US$11,500). After the Aygo and the Etios, Toyota will urgently need to develop further low-cost models if it wants to generate growth and cash flow in the emerging markets that will help it get back to pre-crisis levels. The Japanese company is not lacking in experience in this price segment, having considered the motorization of the emerging markets – and the development of low-cost, robust, entry-level models – to have been a top priority for many years. Furthermore, it has a very promising majority stake in Daihatsu, but has not yet used this to sell this company's micro-cars in high volumes outside of Japan.

In the race to dominate the car markets around the world, both of the major rivals have staked out territories: VW has a strong position in growth-hungry China, a market with total sales of 18 million vehicles in 2010, around 12 million of which were passenger cars. Toyota is still among the dominating forces in the US market, which recorded roughly 11.5 million new cars and trucks sold in 2010. VW is weak here, and for as long as there is growth in the USA, Toyota wants to use it to keep its factories running at full tilt and help both its own recovery and that of the Japanese economy as a whole. It is in VW's second large market, China, however, that Toyota lags far behind in sales and production.

Nevertheless, the Japanese presence in a range of smaller Asian emerging markets such as Thailand, as well as several Middle Eastern and some South American countries, evens out its weakness in other major markets. VW, meanwhile, is by no means as comfortable on the international stage, with its hopes resting very much on China. With the exception of its ventures into Russia and Brazil, essentially the German group has bet the house primarily on high growth in the People's Republic through 2018.

China is a special case. Its market, based on joint ventures with state-owned enterprises, is not without risks, and the sometimes arbitrary nature of decision making is not without challenges for manufacturers accustomed to largely deregulated home markets. With unusual endurance, VW has shown that it has been capable of achieving its long-term strategies in China over the past two decades, but the huge market has also proved to be both a blessing and a curse. While growth prospects in China may be high, the joint-venture strategy is intended to transfer technology and knowledge to domestic industry with the uncomfortable aim of enabling the home teams, one day, to strike back and aim for global domination.

Much like China for VW, North America has been a growth motor for the Japanese, still having the largest number of Toyota and Lexus customers in the world. Despite the short-term need to offer high discounts in order to overcome the quality issues debacle and to fend off a resurgent GM, the long-term predominance in this market should continue to be advantageous for the company from Toyota City. After the stellar comeback of GM and Ford, the trade dispute that seemed set to unfold has been transferred behind the scenes and is still rumbling on backstage, but no longer right in the media spotlight. The Japan Auto Manufacturers Association has repeatedly begged the Japanese government to intervene in foreign exchange markets to weaken the yen – which of course would make cars built in Japan cheaper in America. A stronger yen, conversely, makes it more difficult for Japanese manufacturers to compete; especially now in the wake of the 2011 earthquake, Japanese manufacturers need all the help they can get. Yet, as expected, the American Automotive Policy Council, which represents Chrysler, Ford, and GM, strongly objects to any such action. Currency seems to be one of the major battlegrounds in the trade dispute between the two big economies, Japan and the USA.

4.7 THE FINAL LAP

Both Volkswagen and Toyota have staked out their turf and are well armed for the upcoming fight for the top spot. Both have armed themselves with individual weapons: China vs. North America, modular toolkits up against the Toyota Production System, efficient combustion engines confronting a hybrid technology. Volkswagen has overcome many of its problems and is making an exceptionally powerful start in the new decade with new ideas, a thoroughly revamped model lineup, and better, more fuel-efficient engines. Clearly, with all its strategies working out better than expected over the last five years, Volkswagen has experienced a sustained period of fortunate business. Its Japanese rival, meanwhile, has been experiencing the polar opposite since the economic slump, the recall crisis, and the natural catastrophe on its home island; however, it is nowhere near as weakened as many had assumed in 2009–10. It has overcome its difficulties and is making a fresh start in corporate culture as well as quality management. Further to this, Toyota is now on an emotionally charged mission to help rebuild the Japanese economy following the 2011 earthquake.

The German VW group will be able to achieve its sales goals for 2018 even earlier if it works hard in the BRIC states and if the world economy continues growing, increasing car sales worldwide. Toyota, too, has come up with fresh ideas and a new vision for the future. Despite having broken with old paradigms, it will focus its unique energy on driving down costs even more consistently in order to keep competitors at bay, from VW to Hyundai. The manufacturer from Toyota City learned early on that efficiency and productivity are the keys to success, and that this is about making processes as efficient as possible, from assembling cars and building factories through to launching new models. This, for Toyota, is the art of carmaking, and there is no doubt that the company has not yet reached the pinnacle of that art.

There are few automotive companies with the size and power needed to challenge Toyota: it almost had to be Volkswagen. The race, which began back in 2007, has drawn neck-and-neck which, above all, is an early success for VW. The audience is now anxiously waiting for promising initiatives from Wolfsburg, which could leave an imprint on car markets worldwide – as VW had done with the Beetle, the microbus, or the Golf. For the next few years at least, VW and Toyota will joust with one another – and neither will stop trying to be ahead of the pack in the fields that decide success in the car market of tomorrow: fuel-saving engines, new business models for green cars, and motorization for growth-hungry emerging markets.

4.8 COMPARING CORPORATE VALUES

Volkswagen and Toyota are two rivals with two very different cultural backgrounds and defined therefore by very different attributes (Figure 4.1).

Japan is by tradition a society strongly oriented toward consensus, mirrored in the company values of the Japanese carmaker: whenever a decision has to be made, everyone involved is brought into the process. Once the decision has been made, however, it is respected by everyone involved and cannot be changed at short notice. Changing a decision is a loss of face for everyone involved in making it, and loss of face is the worst possible embarrassment in Asian societies.

Volkswagen on the other hand is marked by rivalry between brands and managers, guerrilla culture, and heroic mavericks; it is a company of single personalities, not of collectives. Managers are expected to show drive and power, and to demonstrate this frequently in order to win people over. This gives Volkswagen its power to continually reinvent itself and adapt to new challenges.

In Japan, by contrast, continuity is highly prized. As a carmaker Toyota is characterized by persistence, by perseverance in achieving goals once they have been set. This pushes individuals and the importance of who exactly makes decisions down the agenda; it makes personal rivalries unimportant

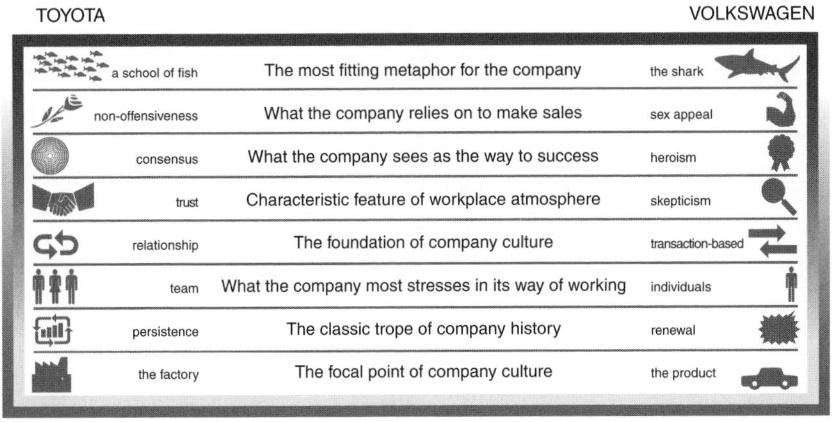

TOYOTA		VOLKSWAGEN
a school of fish	The most fitting metaphor for the company	the shark
non-offensiveness	What the company relies on to make sales	sex appeal
consensus	What the company sees as the way to success	heroism
trust	Characteristic feature of workplace atmosphere	skepticism
relationship	The foundation of company culture	transaction-based
team	What the company most stresses in its way of working	individuals
persistence	The classic trope of company history	renewal
the factory	The focal point of company culture	the product

FIGURE 4.1 **The appendix of corporate values.**
Source: PA Consulting Group.

compared to the need to keep face for the company; and it has allowed Toyota to perfect teamwork as its overriding organizational characteristic. Relationships with suppliers are also founded on mutual trust and long-term benefit for both sides, while short-term advantages for one side or the other are left unexploited.

The difference in culture is even recognizable in the different product philosophies within the two companies. Toyota aims to design cars that no one can dislike, wherever they are sold in the world; whether they actively *like* the cars is relatively unimportant. At the same time, Toyota wants to offer alternative energy vehicles to the mass market, since protection of the environment has been fully integrated into the long-term goals of the company. Volkswagen on the other hand produces dynamic, distinctly masculine designs with an element of style: every model, from compacts up to sedans, is intended to be punchy.

5

EPILOGUE: NEW CHALLENGES AND NEW ENGINES

The automobile? A merely temporary phenomenon; I myself hold with horses.
—Kaiser Wilhelm II of Germany (reigned from 1888 until 1918)

5.1 POPULATION GROWTH AND OIL SUPPLIES

The one-on-one race to the top of the car industry we have been examining thus far will be played out over the next decade, although many of the questions it raises will remain open through 2020 and well beyond. Volkswagen and Toyota are battling each other on some highly interesting fields, seen by many as no less than the fertile ground from which the next industrial revolution is springing. Several issues of crucial importance for humanity as a whole – global population growth, the oil-based world economy, alternative fuels, and the new variety of propulsion technologies – will of course have major consequences for the way in which Volkswagen, Toyota, and the entire industry work.

At the time of writing in 2011, there are already 6.9 billion people on our planet. Thanks mainly to improved provisioning with foodstuffs, progress in medicine, and – a dimension that is sometimes neglected – the widespread use of fossil fuels, this figure has quadrupled since 1900, and, while the rate of population growth has been declining since the 1980s, it is still 1.1 percent a year. This means that the world population is increasing year after year by roughly 80 million, which is almost exactly equivalent to the entire population of Volkswagen's home country Germany or of California, Florida, and Texas combined. Current predictions are that the total human population of planet Earth will grow by almost a half to nearly 10 billion before starting to decline again, with the turning point coming between 2050 and 2100, according to various sources that research average life expectancy figures.

Since the end of the Second World War, not only has the world's population grown, but with it the world's gross national product (GNP); in fact, it has been averaging a growth rate three times higher than worldwide population figures – with the one exception being 2009, the year of the economic crisis, which the *CIA World Factbook* lists as the first in 60 years in which global GNP shrunk.[35] Nevertheless, the average high level of growth is allowing increasingly large numbers of people to participate in global wealth, especially in the dynamic BRIC states (Brazil, Russia, India, China). These countries have long been looking for ways to improve the prospects of their inhabitants, especially those in the rapidly-growing cities. Worldwide urbanization is expanding, with 180,000 moving from rural to metropolitan areas every day. The number of cities with over a million inhabitants has doubled in the last 30 years to 400, and the number of megacities (defined as urban agglomerates with over 10 million inhabitants) is growing disproportionately fast at 4 percent per year. The global population is slowly becoming a global city (Figure 5.1).

This is good news for automotive manufacturers. Megacities are enormous potential markets, with an estimated three-quarters of global traffic miles being driven in these cities by 2050. Moreover, given continued global economic growth over the next 40 years, more than 2 billion people will, for the first time ever, be able to afford a car. This increased purchasing power will lead to huge investments in streets and essential services such as water and

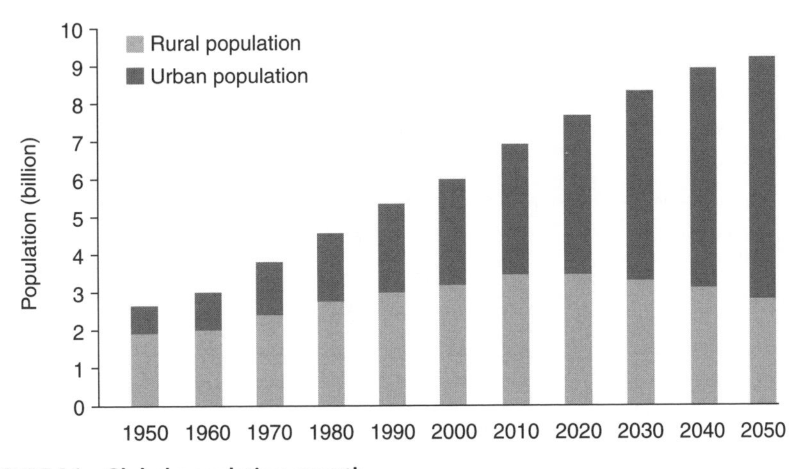

FIGURE 5.1 **Global population growth.**

Source: www.un.org, United Nations 2010.

waste disposal, public transport, and (not least) individual motorized transportation. By the same token, however, this will lead to a correspondingly large increase in energy consumption, with the International Energy Agency predicting an average increase in worldwide oil demand of 0.6 percent annually. Furthermore, energy use in Asia will overtake that of North America, currently by far the thirstiest continent (Figure 5.2).

Yet oil is a finite resource, and the exponential increase in users of this fuel will see oil reserves dwindling faster and faster. Even the recent spate of large oil finds will not be sufficient to reverse this trend.[36]

In order to quench the worldwide thirst for oil, producers are using ever more expensive equipment to find and exploit new sources – and they have no small degree of success to show for it. Discoveries in 2009, especially deep-sea discoveries, were numerous and large, but fraught with risk; however, no amount of new oil can change the bald fact that oil is not an unlimited resource. This became obvious once more with BP's disaster in the Gulf of Mexico. Risky, new oil supplies do not resolve the core issue. Even the biggest of the new oil fields, discovered in 2009 – the Brazilian Tupi field with over 8 billion barrels – would be just enough to meet 100 days' demand at current worldwide usage rates of 80 to 85 million barrels per day. At the same time, production at the world's 500 biggest oil fields – most of them in the Middle East – is dropping by 9 percent per year.[37] Oil is running dry, despite advanced methods for pumping it.

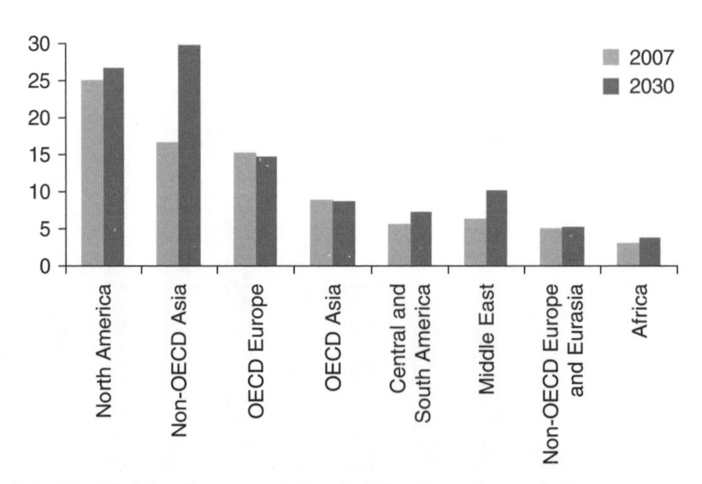

FIGURE 5.2 **Worldwide oil consumption (million barrels per day).**

Source: *Annual Energy Outlook 2010*, Early Release Overview (Reference case) December 2009, Energy Information Administration, US Department of Energy.

The consequence of this is that at some point oil as a raw material will become very scarce, and this fact will most likely lead to further speculative bubbles, price fluctuations, and perhaps even oil wars; because as surely as oil reserves will continue to decline, economic growth will continue to send new car drivers to the gas stations (Figure 5.3).[38]

With supply and demand threatening to diverge severely, not to mention the increasingly dramatic problems caused by CO_2 emissions, the time available for developing alternatives to oil is alarmingly tight. This is nothing new, and for decades now, research has been conducted into alternative energy sources; yet very few are technically and economically feasible.

5.2 BIOFUELS: GREEN GASOLINE

Biodiesel, bioethanol, and other new, biosynthetic fuels: all of these are the concrete results of the dream of CO_2-free, sustainable gasoline – and this is a dream which is as old as the gasoline-fueled automobile itself. Henry Ford, for example, conceived his famous Model T as a vehicle that would run on alcohol of agricultural origin, which he considered to be the fuel of the future. His idea for the Tin Lizzy – a colloquial term for the Model T – was to get America

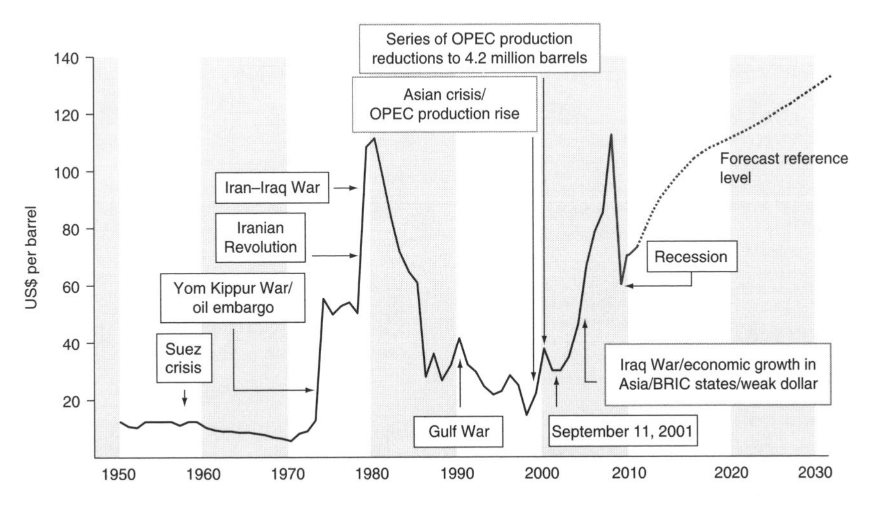

FIGURE 5.3 **Oil price curve based on yearly averages, 1950–83 Arabian Light, 1984–2008 Brent dated, US average from 2009 on (adjusted for inflation).**

Sources: 1950–2008 *BP Statistical Review of World Energy*, June 2009; from 2009 on, *Annual Energy Outlook 2010*, Early Release Overview (Reference case) December 2009, Energy Information Administration, US Department of Energy.

moving, and to do the same for America's agricultural industry which was to produce the alcohol to fuel it.[39]

How wrong Ford was! Due to its easy availability, low price, and the reliable refinement and supply brought in by John D. Rockefeller's Standard Oil Company, gasoline was to become America's key fuel. Nonetheless, it has taken almost a hundred years for the notion to catch on that plants can be used just as well as crude oil to make thoroughly viable liquid fuels. The dream thought dead has been revived (Figure 5.4).

First-generation organic fuels

Biodiesel is the most common biofuel in Europe, where it tends to be made from rape seed and sunflower seed. Whilst it remains an emerging product in the USA, biodiesel is already well established on the European continent, in Germany above all, where the government put an early emphasis on renewable fuels and passed legislation to this effect at the turn of the millennium; an ever-increasing proportion of biodiesel must be added to standard diesel fuel, with the figure set for 2010 being 5.75 percent. In order to get this initiative moving forward, the German government freed biodiesel from oil tax through to 2008.

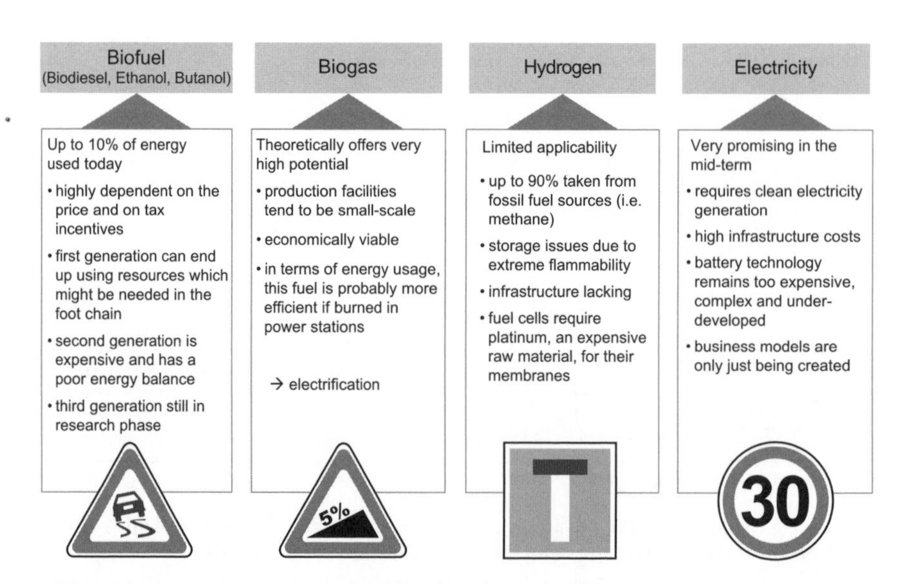

FIGURE 5.4 **Alternative fuels: an overview.**

Source: PA Consulting Group.

Nevertheless, the economic bottom line for biodiesel has remained problematic. Instead of powering Europe's largest car market into a new era of clean, organic fuels, many of the biodiesel refineries opened just a few years ago are already facing bankruptcy – not just in Germany, in fact. Furthermore, there is now serious ecological concern being voiced by researchers, who point out that biodiesel's environmental effect may not be as benign as first thought.[40] Aside from oilseed crops, other oily plants such as sugar cane, corn, and other grains can also be used to produce "green" biofuels, with fermentation of sugars and starches producing bioethanol/bioalcohol. Ethanol, also known as pure or drinking alcohol, has been known to humanity since as early as 900 AD, when it was produced by Persian wine-makers as a by-product. Despite the fact that it has different chemical properties to gasoline, the two can be mixed and used as fuel; in engines that have been adapted to take it, ethanol can even be used alone as a fuel, it simply needs to be refined so that its water content is removed.[41] Already there is no shortage of flexible-fuel vehicles built to run on the mix and, even without adaptation, conventional cars can take up to 5 percent ethanol mixed with petrol; this could make a considerable contribution to capping CO_2 emissions.

The agricultural giant Brazil is the country which has made the most of bioethanol, produced using sugar cane, with the gasoline sold at stations there containing roughly 26 percent ethanol. France, another major agricultural player, is mixing up to 10 percent ethanol into its super-unleaded and other countries are following suit by introducing E10, a similar mixture of 90 percent conventional gasoline with 10 percent ethanol. Nevertheless, rules apply with regards to the use of bioethanol, and due to the corrosive effect it has on seals, engines have to be adapted to burn more than a certain proportion of bioethanol.

Many car manufacturers are adapting to this, too. As a biofuel pioneer Volkswagen is now offering cars that can run on any form of gasoline-ethanol mix. This is achieved by the use of an alcohol sensor which registers the exact mix of gasoline and bioalcohol that has been put into the tank and then adapts the fuel use accordingly. In 2003, Volkswagen was the first manufacturer to offer this "flex-fuel" technology (produced by the automotive supplier Bosch) in the Golf 1.6 Total Flex for the Brazilian market. Now, there are roughly 2 million such Volkswagen vehicles in circulation and all manufacturers have followed suit in offering the technology.

In Europe and the USA, meanwhile, and in some developing countries, bioethanol is produced using grain, corn, and sugar beets, which require the use of valuable agricultural land that could otherwise be used for producing food – a fact that did not go unnoticed by the scientific community and the media. The explosion of oil prices between 2003 and 2007 and the ensuing high demand for bioethanol then led Brazil to massively expand sugar cane cultivation at the expense of rainforests, and bioethanol was suddenly seen as a cause of rainforest destruction, hunger, and suffering. The mood turned, and as the expansion of bioethanol became politically unpopular, the practice was brought to a halt in Brazil due to a broad political consensus.[42]

The major problem with first-generation biofuels is that they rely on oil seeds or fermentable sugars and starches. There is a good reason why humans rely on precisely these products as the staples of their diet, given the high energy-density in the materials in question. Therefore, first-generation biofuels end up competing with humans for their primary source of calories. Nevertheless, on the way from fields either onto plates or into fuel tanks around the world the energy contained in the leaves and stems of seed crops – usually indigestible – is lost as it stays on the fields and rots unused. As the parts of the plants used are separated from a lot of unusable biomass, biofuels that base their production on this indigestible plant matter would of course be a far better alternative. This is what next-generation fuels are trying to achieve.

Second-generation organic fuels

It is precisely this that is the rationale for second-generation biofuels which, in simple terms, carry out an accelerated version of the chemical processes which, over many millennia, turned plant mass into crude oil. These fuels take cellulose as their basis, which is found in stems, leaves, straw, wood, and even organic waste. Theoretically, this way of using waste products increases the productivity of farming land, but the amount of effort required to release the sugar contained in tightly-packed cellulose is unfortunately very high. The processes used for second-generation fuels include the use of biotechnical methods such as enzymes, which essentially mimic cows by "digesting" cellulose into glucose, and of biomass-to-liquid techniques, which can turn almost any biomass with carbon into a synthetic carbon monoxide-hydrogen gas. This gas is then liquefied using the Fischer-Tropsch process, developed 80 years ago to turn coal into a liquid fuel.

Turning biomass into gas requires extremely high temperatures (above 1000°C or 1830°F), and therefore large, durable, and very expensive facilities if the processes are to be kept efficient. Such facilities, however, are only now being tested – with very variable results to date. Currently, fuel produced with these methods is not competitive with oil; but this may well change in the future.

A pioneer in biomass-to-liquid processes is the German company Choren Industries which runs a test plant near Freiberg in Saxony. The first three letters of the company stand for carbon, hydrogen, and oxygen, the three building blocks of all organic life and therefore of all conventional fuels. The last three letters are an abbreviation of the word "renewable".

According to the company, their facilities can produce up to 18 million liters of synthetic diesel per year using waste wood. Volkswagen, Daimler, and Shell all held minor stakes in the company, which suffered from the enormous difficulties to scale the plant from laboratory to mass production conditions and went into insolvency mid-2011. Volkswagen is also cooperating with the Canadian company Iogen, a market leader in cellulose-based ethanol produced from straw. Through its choice of partners here, Volkswagen shows that it is very much capable of anticipating long-term trends and getting involved in all stages from research through to industrial production. Yet it will take some years to see whether this process will ever be mass-market ready.

Experts calculate yearly capacity per hectare as roughly 4000 liters (1050 gallons), which would represent a tripling of volume compared with rape-seed oil and a doubling of ethanol capacity. Theoretically, at least, biomass-to-liquid fuels have the kind of efficiency required for the mass market, but there is as yet no proof that they can be produced on this scale. What is certain, however, is that, for quite some time yet, conventional oil will remain cheaper.

Third-generation biofuels

If second-generation biofuels sound futuristic, the third generation being developed on the basis of algae is like something out of *Star Trek*. Micro-algae are the world's fastest-growing biomass, and some species contain up to 80 percent oil. This has long been known in the scientific community and, as early as 1978, President Jimmy Carter founded the Aquatic Species Program to systematically investigate the energy potential of algae in the wake of the oil crises of the 1970s. The program was canceled in 1998, however, when the oil price sank so low that this form of fuel production was seen as economically unviable, despite initial successes.

A few things have changed, however: fuel prices have risen, and genetic manipulation technology will allow new approaches to cultivating algae. Similarly, genetically modified yeasts may be used to turn entire plants into biotubanol (another futuristic biofuel to watch). The average amount of energy that can be gained using algae outside laboratory conditions, however, remains completely unclear, and there are concerns at economic feasibility. This puts third-generation biofuels well beyond our reach for many years.

5.3 A TANK FULL OF GAS

An interesting organic alternative to gasoline could be the use of methane made from plant matter for vehicles which run on natural gas. Most experts, however, consider the fact that methane is a potent greenhouse gas makes it a climate risk and do not see it as suitable for a car fuel source. Many experts also think that methane should be used to produce electricity right at its source of production rather than being compressed or liquefied, to avoid costly transportation and reduce emissions. This would allow methane to contribute to emissions-neutral electricity production, but would rule it out as a fuel.

The key advantage of biogas is that the plant matter harvested to produce it does not need to be dried or cleaned, and that biogas facilities can use many types of plant as raw material. In a process similar to what goes on in the stomach of a cow, the crop is left in tanks to be anaerobically digested, producing methane which can supplement natural gas drawn from fossil fuel extraction and be used for vehicles already running on gas.[43] Furthermore, by-products of the biodegradation process can be used to fertilize fields.

5.4 HOPING FOR HYDROGEN

Most hydrogen-powered concept vehicles currently being tested on the road are not powered directly by hydrogen, but actually fueled by electricity.[44] The electricity, however, is not stored in a battery, but comes from a fuel cell and is generated as hydrogen and oxygen react with one another – essentially the opposite of electrolysis.[45] The only direct by-product is water, which escapes as water vapor through the exhaust; this makes hydrogen, environmentally speaking, cleaner than a whistle.

Another aspect of the hydrogen vehicle which gives it an advantage over electric battery-driven cars is that fuel-cell technology gives it a far greater range. The major disadvantage, however, is that hydrogen needs to be stored either at very low temperatures or under very high pressure. Furthermore, hydrogen atoms are so small that they are quick to escape, either through tank caps, or even through the walls. As astonishing as it may seem, hydrogen can leak through solid steel, meaning that a full tank of hydrogen is empty after three months at the latest.

It is not just hydrogen storage which turns out to be tricky; producing hydrogen is challenging too. Due to its high reactivity, hydrogen almost never occurs naturally, and must therefore be artificially prepared, but this requires a lot of energy. Theoretically, it would be possible to produce hydrogen by reverse-electrolysis and power the process with renewable energy (sun, wind) but this process is comparatively expensive and has a low energy efficiency.[46] The current method of gaining hydrogen in industrial quantities is by steam-reforming natural gas and, as long as the natural gas used is taken from fossil fuel sources (as is usually the case at present), hydrogen as a whole is not sustainable.[47] Other factors that make hydrogen seem less suitable as a mass-market fuel are the new network of hydrogen stations that would be necessary and the high cost of platinum, which is required for the fuel-cell membranes.

5.5 ELECTRIC CARS: THE DAY *AFTER* TOMORROW

In view of the considerable problems with other alternative fuels at present, all hopes for the future of the car rest on electro-mobility. In fact, it seems difficult to imagine a future in which we will not be driving electric cars, and what many members of the public do not understand is why we are not already doing so.

"After all," goes the argument, "we already have relatively cheap, high-performance battery technology – such as the lithium-ion batteries in mobile phones and laptops – so why did we have to wait for a sports-car manufacturer like Tesla to come along and put the things in a car?"[48] Yet it is not as if this has not been tried before: indeed, it has been tried several times, with varying degrees of success.

The electric rollercoaster

In fact, the first electric car predates the gas-driven automobile by a good 50 years, with Robert Anderson of Aberdeen in Scotland building the first one in 1839; electric motors then remained the primary form of propulsion for smaller vehicles for several decades.

The early years of the twentieth century, however, saw the development of reliable starter motors, which was a great step forward for gasoline-powered vehicles, as well a less technically revolutionary but equally important accomplishment: high-speed racing cars fueled by gasoline. Due to their heavy batteries, electric vehicles could not keep pace, especially over longer distances, and the gasoline engine won the day.[49] There were nevertheless various attempts throughout the twentieth century to revive the electric car, but these received no priority at all until the oil crisis of 1973–4 and the growing ecological consciousness of the 1980s gave the matter a new urgency. So pressing was the topic of the electric car that the US government financed a prototype in the late 1970s. Yet after NASA engineers had tested the vehicle, their verdict was withering: too unreliable, too expensive, and far too short range. This concept car ended up in the car crusher. Nevertheless, bad smog and the growing ecological movement led to increased public demand for cleaner forms of energy, both in the USA and in Europe. A signal for this change of direction was California's new limits on harmful substances in fuels, which came into effect from 1970 onwards. Since these limits could only be reached by installing expensive catalytic converters, there were clashes with the car industry, which was anxious over whether it would be able to pass on this cost to customers. However, the state continued on its course, with the California Clean Air Act of 1990 going so far as to specify that 10 percent of all new cars in the state would have to be emissions-free as of 2003. California intended this to force manufacturers to prioritize hydrogen and electric cars and, despite loosening up the legislative measures subsequently, the political pressure was not completely ineffectual, with General Motors and several other manufacturers putting electric cars that were close to mass production on the road in and after 1996.

The GM EV1 – An electric legend

The best-known of the mass production-ready electric vehicles was GM's EV1. The battery system was based on a 500 kg (1100 lbs) lead-acid battery which accounted for about half the weight of the vehicle. After 3.5 hours charging (at a 220V line), the vehicle had a range of 110–150 km (68–93 mi). GM put the development and production costs at over US$1 billion, and this meant that the car was expensive, with a three-year lease costing over US$30,000 at that time.

Sales were poor. After putting just 800 on the road, the experiment was cancelled. With leasing customers, GM had reserved the right to recall the vehicles after the three-year test period and scrap them; this was done owing to concerns about safety and because there was no further production of spare parts. Today, just a handful of EV1s are left.

Following the demise of the EV1, legend was added to the tragic story. Environmental campaigners accused the car industry of deliberately sabotaging the development of electric cars. There was also suspicion of the oil industry, which bought many of the patents from the failed electric models. In 2006, a documentary was released under the title *Who Killed the Electric Car?* in which GM's decision to cancel the EV1 is critically reevaluated. This film duly gave birth to a variety of conspiracy theories. In spring 2011, Chris Paine, the director of the film, released a sequel to his documentary success: *Revenge of the Electric Car.*

Meanwhile, tests elsewhere in the world provided no less mixed results. In Germany, the government financed a large electro-mobility project on the Baltic island of Rügen. From 1992 onwards, around 60 electric cars produced by five manufacturers were authorized for testing on the island and, here too, the main problems were battery-related: too expensive, too unreliable, too heavy, too little capacity. What also became clear on Rügen was that electro-mobility is not necessarily more climate-neutral than gasoline-powered transportation because, in all likelihood, generating the electricity used will have involved considerable CO_2 emissions itself.

Just how clean are electric cars?

This is a question that should be asked more often, since there is little to be gained with regards to climate change by burning coal in power stations instead of oil in car engines. In fact, the former represents a quadrupling of CO_2 emissions compared to an average gasoline engine. This shows us that the decisive factor for calculating the environmental advantage of electricity is the energy-generation mix used to produce the current being used to drive; electricity is produced using everything from water power and nuclear reactors through to fossil fuels such as coal and gas. In many markets, customers can already choose which of these generation methods they wish to finance and the climactic effects of these choices are widely varied.

Here is a thought experiment to prove the point.

Let us imagine that Mr E-Power drives 11,000 km (6835 miles) a year using a very economical conventional car in the four-liter category (59 mpg). In order to power his vehicle, he needs between 450 and 480 liters (119 and 127 gallons) of diesel per year. To cover the same distance in an electric car, he would only require 1500 kwh a year, as electric motors are more efficient. If this electric energy were produced using an average electricity mix (some would come from renewable, some from gas-powered, some from coal-powered electricity plants), by charging his battery Mr E-Power generates carbon emissions of 910 to 930 kg. By switching from his conventional to an electric car he would reduce emissions by 180 to 200 kg CO_2 – or 17 percent – per year, and this will not be enough to stop climate change. It makes little ecological sense.

If, however, Mr E-Power drives the same annual distance in a less efficient model – e.g. 7.5 liter (31 mpg) consumption – and then switches to an electric car which he powers with nothing but ecologically generated, renewable energy, he reduces his carbon emissions from 1900 kg down to 0. This is impressive, but even then the savings achieved come at a very high cost compared to other carbon-saving methods. If Mr E-Power were able to compare or even "sell" his carbon savings as "emission rights" on the market, he would be disappointed. Current emissions-trading schemes in force in Europe would not come close to covering the additional purchase price of the expensive electric car: his two metric tons of saved emissions would net him about €30 (US$43).

Another valid question is whether there is enough electricity available to enable everyone to drive an electric vehicle? This is not as problematic, however, as the battery question. The high efficiency of electric motors means that the amount of electricity needed is in fact very low. Even a fleet of 1 million electric cars – as penciled in to its national plan for electro-mobility by Germany, the EU's most populous country and Europe's largest car market – would not pose any particular problems for electricity generators. To supply 1 million electric vehicles with zero-emissions energy would take no more than a wind farm with 500 standard wind turbines. By way of comparison: in 2007 alone, Germany erected around 1700 turbines nationwide, while the Roscoe Wind Farm in Texas (currently the world's largest) contains 627 – or roughly 25 percent more than needed for a fleet of 1 million cars. Based on these calculations, it becomes clear that there is no issue (apart from somewhat higher capital – and therefore energy – costs) with green electricity production. Furthermore, there is a clear security and defense policy aspect to this, in that wind power cannot be squeezed, sabotaged, or blockaded. Public interest is therefore growing.

While interest is growing – albeit from a very low starting point – the pressure on car manufacturers to produce an electric car has reached high levels, and the entire sector is reacting with uncertainty and streams of new announcements. Nowadays, barely a motor show passes without some kind of innovative, high-profile electric vehicle. At the time of writing, we are caught up in a climate of electro-mobility hype, but this should run out of steam in late 2011 as the concrete prospects loom larger (Figure 5.5).

The road to mass electric motorization is a long one, and several issues related to material physics and battery chemistry cannot simply be waved away, even with the magic wand of subsidies. What are these issues precisely?

Electric cars are different

Electric cars take the energy they need to propel themselves from batteries, and these batteries are charged on normal domestic power supplies. At the

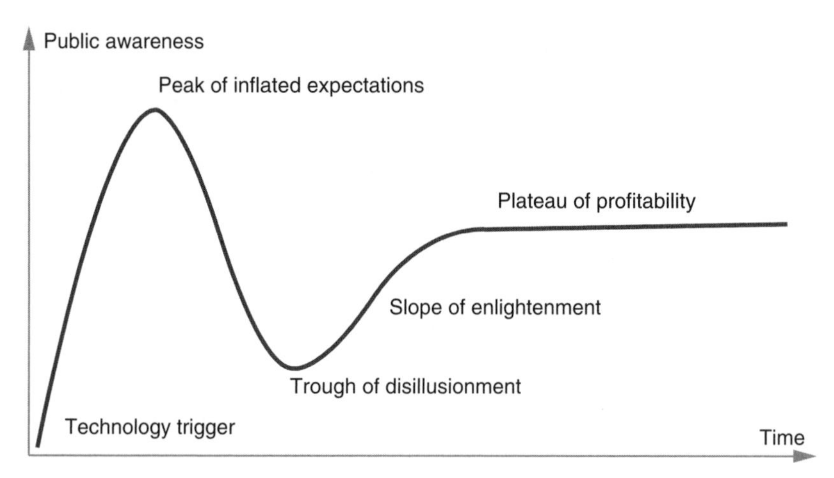

FIGURE 5.5 **Technology hype cycle**
Source: Gartner Inc. 2010.

current level of technological development, large and heavy batteries are required in order to power vehicles.

Example: The Volkswagen E-Up

Volkswagen is currently working on an electric car derived from the New Small Family, the market entry planned for 2013. Conceived as a vehicle purely for city use (with space for three adults and one child), it weighs 1085 kg (2.390 lbs), of which just under a quarter (240 kg or 530 lbs) is battery weight. The battery is a lithium-ion model producing 18 kwh that can be charged at home and, depending on the strength and voltage, can be recharged in as little as an hour. It has a range of around 130 km (80 mi), and the engine is intended to perform at 40 kw.

Due to the heavy battery, and in order to keep their range, electric vehicles will not be able to transport goods and will also have a much lower top speed than gasoline-driven models. Purely electric cars are simply a different, new category of vehicle that will have to be used differently. This means that it is not entirely certain that in the future all-electric cars will look like today's models at all. Design, number of wheels, business model: everything is up for discussion. Just as tractors have developed differently from passenger cars due to their specific uses, electric vehicles too will develop differently from gasoline-powered passenger transport; it is quite possible that electric vehicles will specialize in covering short trips within towns and cities, i.e. shopping, commuting, and transport to and from schools or other institutions.

Certainly, electric cars will not be able to match gasoline-fueled vehicles for range any time in the foreseeable future. Nevertheless, electric cars must still have three-figure ranges in order not to scare customers off, and after this, electric car buyers will simply have to drive differently. In cities electric cars will certainly offer different possibilities; whilst gasoline transport works on a point-to-point program, electric vehicles will become part of the communications network to help them around the mega-jams of the megacities. These cars will most likely communicate with mobile technologies, may have their own IP addresses, and will be able to plan routes taking account of the driver's preference, transport interchanges, and fast-lanes for electric vehicles. This intelligent navigation and connectivity (essential to all future forms of mobility) will also be used for tolled services (highways, parking), and, of course, will keep the driver apprised of the available range, remaining battery power, and the nearest recharging locations.[50]

Much of this has little to do with the actual engine, of course, but will definitely be brought to the forefront as the advent of electric cars changes driving habits. Furthermore, beside their ecological advantages, electric vehicles offer other technical benefits: electric motors are so quiet in comparison to the gasoline engines we are used to that the vehicles will most likely have to produce artificial noise when traveling near pedestrian walkways. Another advantage is that since electric motors can maintain a consistently high torque over a much wider range of revolutions per minute, gearboxes will be made largely redundant; indeed, the high torque at low revolutions gives electric cars a sporty note when accelerating that should even please the occasional motor enthusiast.[51]

Unfortunately, however, this range of possibilities has hardly been exploited at all in the concepts and test cars of the last few years, which are still essentially linear developments of the horse and cart. But this will not remain the case for long, and once the first truly electric cars are developed, it is certain that designers will make better use of the fact that the center of gravity is much lower (due to battery weight) and that the interior can be fitted out quite differently to that of a gasoline car if desired. This kind of freedom to be different is, in fact, a great window of opportunity for the electric car. Ulrich Hackenberg, Volkswagen's Chief Development Officer, puts it like this: "Perhaps there will be a manufacturer who lands a hit with the first really fashionable electric car – something like Apple did with the iPod and the MP3 player market."

Yet another key advantage of electric motors is their far superior energy utilization. With a classic combustion engine, only about 2 percent of all the fossil fuel energy produced is then used to transport the passenger (see Figure 5.6).

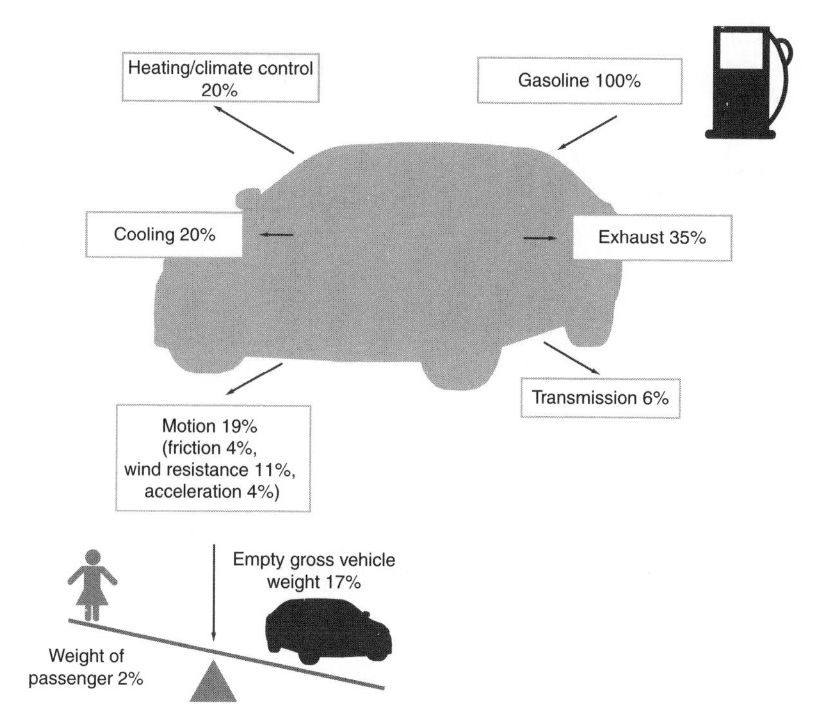

FIGURE 5.6 **Fuel usage distribution in an average vehicle – where does the gas go?**
Source: World Business Council for Sustainable Development, in-house diagram.

Electric engines, however, are far better at sending the energy they have saved down to the road and propelling the vehicle, being three times more effective here than gasoline motors. This becomes clear when the amount of energy required for both types of car is examined.[52]

Unfortunately, however, electric cars are still carrying a serious disadvantage – a massive disadvantage in fact – the batteries. These are set to remain very heavy and very expensive for the foreseeable future (see Figure 5.7).

In order to deliver a satisfactory driving range of 100+ miles, the electric car batteries need to store between 18 and 35 kwh, depending on the weight of the car, engine performance, climate-controlled interiors, etc. For comparison, hybrids tend to run on 1.5–2 kwh batteries. In 2015 these much bigger batteries for all-electric vehicles will still cost between €6000 and 10,000 (US$8600–14,400),[53] before taking into account costly items like the battery-protection crash-box,[54] and performance and recharging electronics. These costs are simply too high for mass-market readiness, and even early-adopting customers are still by and large not ready to part with US$100,000 (€70,000) for a "battery on wheels" like the Tesla.

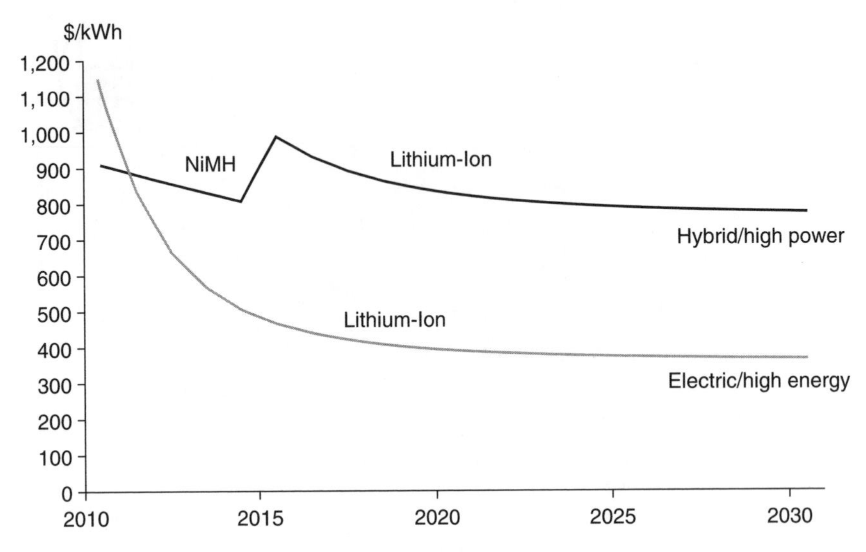

FIGURE 5.7 **Changes in battery costs per kwh (whole battery including cells).**
Source: PA Consulting Group.

The battery is the problem

So batteries are too heavy, their life expectancy and range is low, and they are not yet particularly reliable. Even so, these limitations have been present as long as batteries themselves, meaning that the race for electro-mobility is very much, as VW CEO Martin Winterkorn put it in a speech given in July 2009, "more of a marathon than a sprint." This marathon is well under way, however, with intensive research carried out into batteries since the early 1990s. Japan and China have built up a good lead here after Germany and the USA bowed out around the turn of the millennium. Currently, it is the lithium-ion battery that is leading the field, but there is nothing to rule out surprising new discoveries that would change battery chemistry. Yet it will take at least a decade to move innovations from the lab to the factory floor.

Currently, 99 percent of batteries worldwide are made in Asia, with South Korea, China, and Japan dividing this market amongst them. In recent years, however, battery technology has become a top priority for all industrialized nations, and along with China, Germany, and France, the USA is financing battery research in a big way with the aim of securing national economic interests. Each of these countries is taking these measures not just for environmental reasons, but with the aim of securing jobs in future industry. Many of today's jobs would disappear overnight if electric energy were suddenly to replace combustion engines in the automotive sector, and so it is politically opportune to start producing batteries domestically and to encourage research and development activity at home. Nevertheless, the fact that all these countries want to produce batteries means that, at a worldwide level, enormous amounts of highly subsidized battery production capacity are being built up and, even if electric cars were to come into circulation *en masse* in the near future, there simply will not be enough customers for all these batteries.

Just as was the case with computer memory chips and solar cells a decade ago, it is also quite possible that a technology which today is expensive and scarce could soon be available to excess. This would not lead to heaps of cheap batteries however, but rather to a series of bankruptcies in the sector, which in turn would be disastrous inasmuch as the firms concerned would not only be unable to carry the high costs of their under-used factories, but would also be unable to invest in further research.

Hackenberg, Volkswagen's Chief Development Officer shares this view: "Battery technology is only just in its infancy, but we can also expect huge leaps in electric traction," he believes. Therefore, massive investments in batteries at the moment go into developing products that use today's technology, although almost certainly this technology will soon be obsolete and it may prove difficult and costly to retool – extra investment in the hundreds of millions could be necessary – and this would instantly reduce the market to a few, highly-capitalized companies.

There is no one who sees the danger of battery-related dead-ends more clearly than Toyota, who invested heavily in nickel-metal hydrides. These turned out to have little future in electric traction after 2013 because of the lithium-ion successor, and while Toyota can still use them in the short term, there will be some hefty write-downs on research into the obsolete technology in the future. Now the company is faced with trying to jump onto the lithium-ion train as it gathers speed.

Not only is this new technology financially risky, it has other less attractive sides as well. It has a very short lifespan, and whilst customers do not think twice about throwing away rechargeable batteries for their TV remotes after charging them 500 times, this will not be popular with the kinds of battery installed in cars. Even reduced usage does little to help, and if they are ever completely run down, this too caps their capacity and shortens their lifespan.[55]

Another very serious problem is safety. It was not all that long ago that pictures of burning computers and exploding cellphones went coursing through the media by way of warning against lithium-ion technology, and since car batteries store far larger quantities of energy than the average laptop computer, production defects or handling errors by drivers could have horrendous consequences. While it is likely that developers will surmount these battery problems more quickly that many observers expect, there are as yet no long-term test results of the kind that are usual in the car industry.[56]

Until recently, the key to electrification was seen to lie in making vehicles lighter: every pound saved on weight, so went the logic, would yield an extra mile in range. Yet what car makers are only now discovering is that a big difference can be made by recapturing and recycling the energy lost in braking. Magna, for example, is counting on improving the regenerative braking system already used in hybrids and wants to then use it even where, until now, the engine brake absorbed the supplementary energy (for instance, when rolling downhill). The manufacturer demonstrated – using a test vehicle – that early regenerative braking (occurring as soon as the accelerator is allowed to rise above the middle position) can lead to greatly increased range for electric vehicles. If this technique is consistently employed, then vehicle weight becomes essentially secondary. The focus is then transferred to potential energy differentials, friction and air resistance, and losses due to battery cooling or non-essential energy use.

Despite its promising nature however, this technology is still some years away from being ready for mass production; but if it works as planned, electric vehicles in the future will be able to be far heavier than assumed to date. Essentially, rather than losing energy when braking, the car will simply be pushing the same energy back and forth between the motor and the brake.

Cheap electric micro-cars

Today, in China, a fleet of 80 million electric bicycles is on the roads day in, day out. These machines are quiet, produce no direct emissions and, whenever the battery is empty it is unclipped and taken inside to be loaded in the rider's apartment or garage. The technology involved is extremely simple, and all aspects of construction are kept to a strict cost minimum; and just as the ubiquitous cheap motorbikes in India gave birth to the Tata Nano – an entry-level model for India's first-time car buyers often termed a "motorbike with a roof"[57] – it cannot be long before Chinese electric bicycles spawn mini electric cars.

The target customers have very little in common with American and European car buyers, who purchase cars that allow them to go hundreds of miles on the highways if need be. Potential micro-electric buyers in China (and other developing economies) are perfect customers for basic vehicles with a short range produced as cheaply as possible. Electric bicycles have accustomed a generation of customers to batteries, ranges, etc. – in fact, they know nothing else. Nor are they interested as much in crash-test safety, after-sales service, second-hand value, and the other trappings of the Western car market. This robs manufacturers of any criteria on which to differentiate themselves except price, and so a ruinous price war will be a continuous feature of this new segment. Since the only complicated piece of engineering on these vehicles will be the battery, and these batteries will be low-performance and standardized, entry costs for manufacturers new to this market will be extremely low.[58]

This is one of many as yet untapped fields of demand in electro-mobility that could provide battery makers with the volume they need to manufacture for the mass market. PA Consulting Group expects a growing number of micro-electric vehicles toward the end of this decade.

The long run-up

All in all, the prognosis for a genuinely "new form of car," promised by so many manufacturers when asked about all-electric vehicles, is at first sobering. Real electro-mobility will not even start to be rolled out until after 2015, and the road to it becoming the genuine mass-market alternative to gasoline-driven vehicles will be long and winding. There are simply too many obstacles to be overcome.

For even this small degree of progress to be achieved, subsidies will have to remain at a high level and flow freely into electric R&D for many years to come. Moreover, several megacities will have to make a conscious decision to concentrate on implementing electric passenger vehicles as part of a mobility solution. Even before this, plug-in hybrids will take quite a long time to reach their potential, as well, and the first mass-market all-electric vehicles will not have a range much beyond 30–60 km/37 mi (the Chevy Volt has been announced at 60 km).[59] In profitability terms, meanwhile, PHV will continue to be tricky because of having two motors; but if enough subsidies are available to make the costs of switching more palatable to car buyers, the twin-engined vehicles will really start to spread. The USA has set itself the goal of putting 1 million PHVs on the road by 2015. Yet, despite subsidies, this still seems very ambitious due to the various handicaps on the available design concepts and the charging infrastructure.

Two other factors, however, will lead to increased market penetration for PHVs. First, the car buyer does not have to rethink his or her driving habits: there are no limits on range or speed, and future vehicle generations will offer an ever-increasing battery range to make the swap to all-electric less abrupt. Second, many of the combined engines will, simply put, be fun to drive. The flexibility of both a combustion and electric engine together will offer the kind of performance that could almost compete with sports cars. This may not be a particularly ecological argument, but it will be a substantial dimension when it comes to spreading the plug-in hybrids; and with Toyota wanting to add more emotion to its cars, the full fun potential of the combined engines will become more important still.

PHVs will remain a special case for a while, however, and are unlikely to see sales in significant numbers outside of the USA and Japan, where they will benefit from higher subsidies. Volkswagen, meanwhile, will attempt to use its top-of-the-range Phaeton or high-quality Audi vehicles to establish PHV in the premium segment. PA Consulting Group has carried out a meta-study of potential sales for PHV and electric cars which takes into account the models that have been announced for coming years, their points of market insertion, and other, earlier studies. The sales volumes predicted show the figures both for PHV and purely electric vehicles.

Out of the 700 to 800 million vehicles to be sold between 2010 and 2020, only 30–40 million (cumulative) will be either PHVs or, later, pure electric vehicles, and they will only really start to sell in large numbers towards the end of this decade, meaning that electro-mobility will remain a niche product through 2020.

By 2020 at the latest, however, the number of electric cars will start to increase significantly, due to a wide range of factors. Emissions-related law-making, for one, will make the technical side of combustion engines increasingly expensive, while oil prices will continue to increase. Over the same period, batteries will become cheaper, and by 2020, electric cars will begin to change the face of the world's megacities.

At the electricity pump

"Filling up" an electric car will cost far less than it does with a conventional vehicle, but it will take quite a bit longer. Although the exact length of time will depend on the electricity source used (voltage and wattage) and on the battery in question, there will be no more "just popping down to the gas station," because average charge times will be measured in hours, not minutes or seconds.

In fact, until recently, estimates were running at 12 hours for a normal charge (i.e. charging in such a way as to protect the battery), but there is no shortage of scientists who claim that they can accelerate this process by a large factor by using special batteries that will charge in minutes. Yet this kind of technology is still in the laboratory and many experts are very skeptical about its transferability into everyday usage involving large batteries; quick-charging puts strain on batteries, making them overheat, tire, and perhaps even suffer a short circuit due to physical changes if they are pumped too full too quickly. That is why today, in order to protect batteries, the current is usually cut right down toward the end of the charging process, although this of course has the effect of extending the charging time.

Furthermore, the faster electric cars are to be charged, the more energy has to be pumped through a given section of wiring in a given time; and while this poses no problem at all in terms of theoretical rate of energy transfer, as presently installed the electricity supply grid may well have difficulty absorbing this new demand. National energy grids that are less robust, such as those in the emerging markets and the USA, will require considerable investment if large-scale electro-mobility is to be achieved through home charging. The easiest solution would be a "wall-box",[60] an estimated $1300 (€900) investment that would likely be partly financed by many energy providers, in part because it would enable them to sell more electricity to the household concerned. Yet if entire streets need to be re-cabled or if cities want to start offering public charging stations, the investment will be very costly and not come overnight.

Nevertheless, many energy providers have their eyes fixed on the extra energy they will sell, especially night-time electricity, and are happy to invest for the sake of the new income potential. In Europe, in fact, the big producers are so keen to get going that they have already agreed with the European car manufacturers on a continent-wide standard plug. This is a convenient move: who wants to have to drive armed with 20 adapters just to be able to recharge their car when they get to Grandma's house in the next county? So far, Volkswagen, Toyota, Daimler, BMW, Ford, GM, PSA Peugeot Citroën, Fiat, and Mitsubishi, who between them cover much of the European car market, have all agreed to this standard. Whether this standard really will be kept to, however, remains to be seen.

Once thousands and then millions of electric cars are on the roads, additional possibilities come into focus. Known as vehicle-to-grid technology and patented by Audi in 1982,[61] this system avoids one of the key problems of electro-mobility: storing energy. By linking the vehicle to the grid, car battery capacity is opened up as a buffer for times when there is too much energy on the grid (such as late at night, or if high winds are driving turbines particularly fast). When a lot of energy is in use, car batteries that were charged during the down period feed some of their electricity back into the network. Currently, this is only a theoretical possibility since current batteries would age quickly with this continual dis- and re-charging. Until further research is done, this remains an interesting vision for the future, but not yet a profitable one.

On the topic of profitability and figures, communication between electricity consumers, providers, and electric vehicles and the best method for billing are yet further nagging details that have to be decided upon. Providers are already thinking about it, though, and want to offer a complete system like network providers do in the cellphone sector. The business model would be a kind of mobility subscription, with the provider offering the customer a contract for the batteries and access to the network; the vehicle, like cellphones, would be offered as an extra. This is the kind of new idea that Better Place, a US company based in Palo Alto, California, wants to push up the agenda. Another interesting aspect of the company's vision is that, instead of having to kill time while the vehicle charges "at the electricity pump," the empty batteries could simply be exchanged for fully-charged ones at a network of stations.

Swapping instead of charging – Better Place

The California-based company is aiming to do nothing less than install a dense network for mass electro-mobility. Customers would buy an electric car (without the battery) from any given manufacturer and Better Place would take care of the batteries.[62] If the customers decide to charge at home in their garage, at work, or at a Better Place charging station, the electricity used would be free to the customer. The company can offer this thanks to contracts with electricity providers; intelligent software would mean that customers only paid per mile actually driven. If they want to drive long distances, they plan their routes to pass by fully automated battery stations where robots swap over the cells. In a way, this is a return to the age of stagecoaches, where tired horses were swapped along the route and allowed to regain their strength in peace while the coach traveled onwards.

There is no small number of cities and countries that are convinced by Better Place's vision, including Israel (testing charging stations since 2008, planning to roll out nationwide in 2011) and Denmark. In the USA, Hawaii and the San Francisco Bay Area are planning on using the system; in neighboring Canada, Ontario is in talks with the company, as is the south-eastern seaboard of Australia. Aiming for success rather than planning for failure, Better Place is staking its lot on its battery approach. Much of the system's future is riding on whether batteries are standardized and exchangeable; if so, the idea could well work. The technology certainly does: at a battery-changing station opened in Yokohama, Japan, the robot managed to swap over the batteries of the test electric vehicle in just 60 seconds. What remains to be seen is whether and how many vehicle producers agree to produce to meet the needs of the project, and this is the Achilles' heel of the entire idea.

Until now, the engine and the powertrain in a vehicle were one of the key characteristics on which the manufacturers built the product identity, and it is hard to see the big automotive companies being happy to part with control of even one component of the drive system, be it just the battery. This is a big danger for Better Place, as is the prospect of the large energy producers being unwilling to leave the company to corner the electro-mobility market.

Two additional considerations make this venture seem even bolder. First, Better Place is planning to accept the financial and technical liability for nothing less than the most expensive part of an electric car – the battery. How will the company respond when important technological changes happen in this field, as will certainly be the case? Each improvement in batteries could see the most expensive part of Better Place's inventory being devalued by quite a margin.

Second, the idea of having additional batteries ready-charged at a network of swap stations is theoretically a very good one, but it makes for a very expensive system in practice. At unit prices of around US$13,000 (€9000), batteries are not exactly bargain-basement items. Nevertheless, Better Place has taken a courageous step toward sustainable, individualized passenger transport and could well have shown the car world the business model for the future.

Utilities and the adoption of electric vehicles (by Arun Mani)

As indicated in previous chapters, the need to reduce carbon emissions is creating a drive towards convergence between the automotive and energy sectors. What does this mean for utilities (electricity suppliers and network businesses)?

Up until now, utilities in North America and most of Europe have had a relatively down-to-earth business model, and their consumer relationships have tended to be rather simple, based around issues such as billing, new connections, and energy delivery. EVs unlock the potential of very different and more intimate consumer interactions. While this brings challenges, it also provides opportunities to develop new services and ultimately, enable the growth of the EV market. To be successful, utilities will need to get three things right: to shape their strategy around consumer behavior; to build flexible infrastructure that can adapt to evolving technology and consumer needs; and finally to develop a commercial charging model that works for both the utility and consumers. This will require a careful and deliberate approach as there are significant uncertainties surrounding each area.

Shaping a strategy around consumer behavior

Consumer demand for EVs will be largely driven by their economic attractiveness and perceptions of how the individual product will fit consumers' lifestyles. The utilities of the future can impact both variables, and it is in their interest to do so. So while in the short term the impact of EVs on the grid is unlikely to be significant, utilities will need to forge a long-term business strategy to influence EV charging patterns and prepare the grid for the clustered demand that electric vehicles will bring (see Figure 5.8).

Utilities need a sophisticated understanding of consumer behaviour and how it might affect the take up of electric vehicles and their consequent impact on the grid. Certain utilities, like those in the USA, have an advantage as they have a wide range of consumer data already available to them

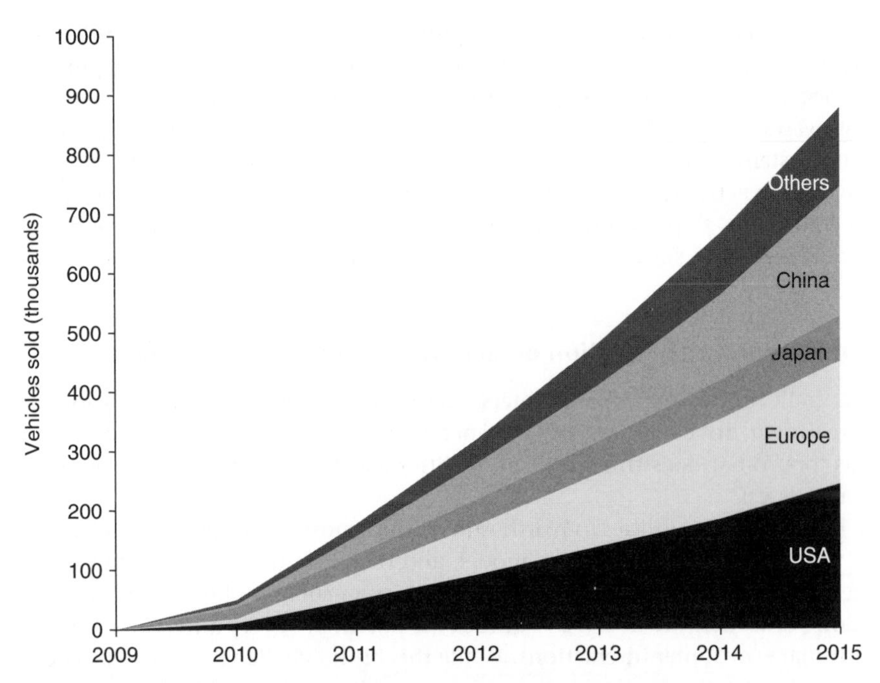

FIGURE 5.8 **Global sales forecasts for plug-in-hybrid and electric vehicles**
Source: PA Consulting Group.

due to the integrated nature of their structure, whereas others, like those in the UK – where the energy supplier is different from the energy delivery company – will need to collaborate with their power suppliers to access these details. This data can then inform further research about how many electric vehicles will be on the road in their service territory, where public charging stations should be located, and when, where and how often, EVs are likely to be charged.

Taking this approach and becoming a customer energy advisor helps to position the utility as an expert in the field, informs the utility about network infrastructure updates required in support of EV's and also promotes the growth of the EV market through educating customers.

Building flexible infrastructure that can adapt to evolving technology and consumer needs

Though utilities can expect to access additional revenue streams through the widespread adoption of EVs, the extent of this will depend heavily on supporting infrastructure being developed. A number of utilities have decided to lead the debate about the development of infrastructure and we are seeing some common issues emerge that they are now working to overcome. In particular, many of these utilities are working to find the right balance between investing in home charging, alongside public charging stations. Industry research shows that the majority of EV owners are expected to charge their vehicle at home rather than in between journeys at public charging stations. However, while public EV chargers are currently more expensive, they are seen as a necessary part of the required infrastructure to support mass adoption and reassure consumers looking to adopt that an EV will be as reliable as a traditional vehicle, even when it comes to long journeys. In the EV world, this is known as "overcoming range anxiety."

There are considerable challenges here, not least in ensuring compatibility of billing arrangements between different regions or countries, dealing with variable battery life, and handling costs. One way for utilities to address both of these issues is to work closely with the automotive sector. For example, Portland General, a US utility in the Pacific Northwest, has teamed up with Ford to share information on charging requirements. The partnership includes working with state and local governments to support charging station permitting, EV tax credits and to develop regulatory frameworks.

Interoperability between systems is another important factor, where different schemes from different operators are being deployed across the country or where drivers cross national boundaries and some interoperable charging systems are gradually being rolled out – such as the UK Plugged-in-Places scheme, a government subsidized program of infrastructure roll out.

Developing a commercial charging model that works for the utility and the consumer

Utilities must establish whether a time-of-use, flat-rate, or fee-based charging model (or a combination of these) best meets the needs of those consumers and their business. And they also need to decide how the cost for EV charging will appear on the electricity bill; for example, does an EV charge like any other appliance or is it shown as a separate line item? And they need to find commercial offerings that will provide a viable proposition for home charging and public charging stations. As home charging units are significantly less expensive – both in up-front unit costs as well as permitting and retrofitting costs – than higher powered public or commercial grade chargers, home charging is expected to be commercially viable. The cost range varies with the types of charging points in play – from below US$5000 for standard charging at home via US$5000–8000 for fast charging at supermarkets up to US$40,000–80,000 for public rapid charging.

These infrastructure questions become more complex in regions or countries with a competitive electricity supply market. For example, the UK is dominated by six major electricity suppliers with networks owned and operated by independent companies. The unanswered questions are therefore very much about who pays for the development of the infrastructure and how. This is further complicated in densely populated areas with a high degree of on-street residential parking such as London, which will require a high density of more expensive public charging points.

Some public authorities are already supporting these developments. In the USA, Austin Energy, in a federally funded initiative, is installing 100 public charging stations with a subscription model giving unlimited access to recharging and offering non-subscribers a flat rate of US$2 per hour of charge. This should mean that no driver will ever be more than five miles away from a charging station. As Karl Rabago, Austin Energy's Vice President for Distributed Energy Services says, "research shows that consumers simply will not buy electric vehicles until they are confident that they fit their lifestyle." This underlines again that utilities have a key role to play in making the take up of electric vehicles attractive to consumers by providing the infrastructure to support their use. As utilities and carmakers adapt their business models, develop new partnerships and lead the charge on greening their sectors, we watch as convergence between these two sectors unfolds.

NOTES

1. This is attributed to Taiichi Ohno, one of the key figures behind the Toyota Production System (TPS), and is roughly translated as "go and see." A founding principle of the TPS, the idea behind it is that, in order to truly understand a problem, one must go to the place where it originated (*gemba* in Japanese); only at the site of the problem will it be possible to solve it. *Genchi genbutsu* is an important problem-solving tool in Toyota processes.
2. "At Toyota, we believe the key to making quality products is to develop quality people": one of the key principles of the Toyota company philosophy is that the most important resource Toyota has is its employees, since it is they who determine the quality of the product.
3. Stewart, Thomas A. and Raman, Anand P., "Lessons from Toyota's Long Drive," *Harvard Business Review*, Aug–Sept 2007.
4. Spear, Steven J., Working Paper 02–043 "Just-in-time practice at Toyota: Rules-in-use for building self-diagnostic, adaptive work-systems," 2002.
5. Womack, James P., Jones, Daniel T., and Roos, Daniel, *The machine that changed the world* (New York: Harper Paperbacks, 1991).
6. Quotation taken from Stewart and Raman, "Lessons from Toyota's Long Drive."
7. Much of the information on the role of *kaizen* in this section is informed by Ulrich Jürgens, a researcher at the Berlin Center for Social Studies who published "Why Toyota has stayed so strong for so long" (title translated; only available in German), Wissenschaftszentrum für Sozialforschung, Berlin, 2007.
8. Quotation taken from Stewart and Raman, "Lessons from Toyota's Long Drive."
9. While the carmaker was expanding, it was always able to take workers who had become superfluous at one location and put them to work at another; for a long time, this system of efficiency savings and staff rotation worked very efficiently, and the dramatic scale-back planned in coming years is likely to be the first time that *shojinka* will cause problems.
10. Nevertheless, pay does increase with the number of years worked.
11. Quotation taken from Stewart and Raman, "Lessons from Toyota's Long Drive."
12. J.D. Power and Associates, 2010 *Vehicle Dependability Study.*
13. *Heijunka* is a Toyota-specific word for reducing the effect of fluctuations; much of the information on this comes from Ulrich Jürgens (see note 9 above).
14. Stewart and Raman, "Lessons from Toyota's Long Drive."

15. Figures taken from the Toyota Deutschland GmbH annual report.
16. Acura was coined in 1986 as the US brand name for luxury and sports cars produced by the Japanese Honda company for the North American and Hong Kong markets. In 2004, Acuras came to Mexico, then went to China and Russia in 2006 and 2008 respectively. Honda is planning to introduce it to its home market, Japan, in 2011.
17. "Mercedes and Lexus locked in race for No. 1," *Los Angeles Times*, November 6, 2010, available at http://www.latimes.com/business/ la-fi-autos-luxury-20101106,0,2321506,full.story
18. Quotation taken from Stewart and Raman, "Lessons from Toyota's Long Drive."
19. Entry for "Toyota" in the *World Data Book*, 2010.
20. Toyota Industries is not just a supplier for TMC, but also contains the automatic loom manufacturing business founded by Sakichi Toyoda; it is therefore a core part of the Toyota industrial heritage.
21. Spear, "Just-in-time practice at Toyota."
22. Dyer, J.H. and Ouchi, W.G., "Japanese-Style Business Partnerships: Giving Companies a Competitive Edge", *Sloan Management Review*, Volume 35, Autumn 1993, pp. 51–63; whilst the figures given in this essay are likely to be higher than present day equivalents, the general trend remains the same.
23. Stewart and Raman, "Lessons from Toyota's Long Drive."
24. The following figures are from TMC as at March 31, 2011.
25. All figures supplied by TMC.
26. A major crisis point in this PR disaster came on February 10, 2010, when a former Toyota US lawyer, Dimitrios Biller, sued the carmaker. Speaking to ABC, he claimed that Toyota felt itself to be above American lawmakers because its headquarters are in Japan; he accused the company of lies, deceit, and unethical conduct. This television interview sent shock waves through the country, especially when Biller further claimed that he possessed material which could prove the accusations and incriminate Toyota. Toyota denied that the documents in question could prove anything to this effect and counter-sued. Almost a year to the day, on February 8, 2011, it became known that an out of court settlement had been reached in which Biller was forced to pay Toyota damages for his breach of contractual confidentiality clauses and prohibited from disclosing any further information.
27. Watanabe commented "Toyota must balance incremental improvement with radical reform": Stewart and Raman, "Lessons from Toyota's Long Drive."
28. See for example Toyoda's interview with CNN, available at: http://money.cnn.com/video/news/2011/04/07/n_toyota_ceo_quake.cnnmoney/
29. Rodriguez, Ann and Page, Chris, "A Comparison of Toyota and Honda Hybrid Vehicle Marketing Strategies", Rocky Mountain Institute, 2004.
30. BBC News website, "How hybrid power surprised the car industry", November 16, 2004, available at http://news.bbc.co.uk/2/hi/business/4015831.stm.
31. Nevertheless, as the profitability studies in section on "Electric Cars: The Day *After* Tomorrow" in Chapter 5 of this book show, over the European market as a whole, hybrid vehicles will soon start to make sense for drivers there – despite the higher purchase price, even without sales incentives. This is due to high gasoline prices; in the USA, it will take far longer for a fully hybrid vehicle to become better

value than a purely combustion-engined vehicle over the entire life cycle of the vehicle.

32. Autoblog.com, "Chademo suggests drinking green tea while recharging your electric car", March 15, 2010, available from http://www.autoblog.com.
33. greenmomentum.com, "Tepco joins Japanese Automakers to create global EV charging standard", March 15, 2010, available at http://www.greenmomentum.com/wb3/wb/gm/gm_content?id_content=5324.
34. It is important to avoid confusing FCHVs with combustion-engine hybrids at this juncture; the former are electric cars hybridized with a fuel cell. This means that the electric battery is not just a buffer supply charged by energy gained from braking and used to add pep to the motor, but the central energy unit in the vehicle. Once the battery is exhausted, the fuel cell can be used to replenish it, meaning that fuel-cell hybridization should be understood as a range extender.
35. According to the *CIA World Factbook*, 2009 was the first year since 1946 in which global GNP shrank (minus 1 percent against a yearly average growth of 3.5 percent) – https://www.cia.gov/library/publications/the-world-factbook/index.html (as at June 24, 2011).
36. This is without even broaching the issue of the limited capability of the planet's ecosystem to absorb additional carbon emissions.
37. International Energy Agency figures.
38. Economic growth will also entail increased demand for other oil products (e.g. heating oil, kerosene).
39. Kovarik, Bill, PhD, "Henry Ford, Charles Kettering and the 'Fuel of the Future'," 1998, available at http://www.runet.edu/~wkovarik/papers/fuel.html (as at July 29, 2011).
40. One of the most high-profile critics of biodiesel is Paul Crutzen, Nobel Prize laureate in Chemistry, who argues that fertilizing the extra rape seed needed to produce it produces large amounts of nitrous oxide (N_2O), a potent greenhouse gas. See http://www.rsc.org/chemistryworld/News/2007/September/21090701.asp (as at July 29, 2011).
41. Combustion engine pioneer August von Otto used an ethanol made of potatoes in his prototype motors.
42. A very promising possible source of plant matter for biofuels, however, is the jatropha plant; it grows on low-quality agricultural land to produce an oil that neither humans nor animals can digest.
43. As a car fuel natural gas has been struggling for years to increase its low market share. The sparse tanking station infrastructure, high refitting costs, and comparatively low ranges continue to make gas-powered vehicles unattractive to consumers, despite improved carbon emissions.
44. One exception to this is the German manufacturer BMW, which has motors which are powered directly by hydrogen. BMW has mothballed this project until further notice, however, in order to improve hydrogen tank-storage technology.
45. Electrolysis (which comes from the Greek for "splitting by using electricity") is the process of separating elements from compounds by using electric current. One example is splitting water into hydrogen and oxygen, and fuel cells do the precise opposite of this, combining oxygen and hydrogen into water and producing heat

and electricity as a by-product. Membranes are used to separate the two gases in order to control the reaction.

46. Using cleanly produced electricity directly via batteries is almost beyond a shadow of a doubt the ecologically and economically better strategy.
47. Even when applied to biogas or biomass, the tried-and-tested use of steam reformation (which consists of separating hydrocarbons present in natural gas – light petroleum and methanol – from hydrogen under high temperatures) is only climate-neutral if the energy source for the process is renewable; currently this is not the case.
48. Tesla Motors was founded in California in 2003 with the aim of developing electric cars. The first vehicle, the Tesla Roadster, was designed by the British sports car manufacturer Lotus, with many of the parts coming from the Lotus Elise (including the light aluminium frame); the body is made of carbon fiber. The first 100 Roadsters were sold for US$100,000 apiece in under a month, with series production starting in 2008.
49. There were nevertheless niches in which electric vehicles survived the twentieth century, the most notable being found in Great Britain in the form of the "milk float". Well into the 1990s, these small vehicles were a common sight on British streets, delivering bottled milk from door to door until the continued rise of the supermarkets sent their numbers plunging. Today, only a few survive.
50. The new Nissan Leaf is already being hailed as a pioneer in terms of vehicle networking in the electric era. Nissan is already testing iPhone interfaces that allow the user to stay in constant contact with the vehicle. Currently, very few features are on offer (programming climate control, checking battery levels), yet the idea of linking the car with a mobile telephony device shows how electric cars will be integrated into the virtual environment of the megacities of the future. All vehicles will have their own internet address as standard.
51. The Tesla Roadster, for example, weighing just over a tonne and going from 0 to 60 mph in under four seconds, is enough to show that electric cars can more than compete with gasoline models in terms of sportiness.
52. Compare with calculations regarding hybrid vehicles in Chapter 2 of this work (see Figure 2.4 and surrounding text).
53. Lithium-ion batteries have very different price tags depending on whether they are being used in hybrid or electric vehicles.
54. Crash-boxes are required so that the battery can withstand serious accidents without catching fire and exploding.
55. There is no shortage of plans to use electronics to allow batteries to be charged up to 500,000 times (this works in theory, at least); yet there is of course no practical experience of this as yet.
56. Many security requirements for automobiles – such as crash-safe battery-cell construction – add even more weight to already heavy electric vehicles.
57. Although it seats four, the Nano is based on motorcycle technology, leading to a much-reduced life-expectancy for many components, as well as a higher rate of accidents.
58. To be more precise, innovations and benefits from improvement will come from cell chemistry and cell construction; the way in which cells are packaged into

batteries and the electronics used will almost certainly not be long-term USPs for manufacturers. There will probably be severe pressure on costs here, too.

59. The design behind the Chevy Volt is called a "serial hybrid"; this is because the combustion engine drives the generator which then charges the batteries. The batteries are the only source of energy for the motor, with the combustion engine having no direct contact with the wheels. This allows it to be run at a constant speed, maximizing efficiency and making for an uncomplicated engine structure.
60. A wall-box is a charging station with a heavy-duty charging point; it has increased safety technology and can be controlled by reference to electricity prices (i.e. only charging at night/below a certain price threshold).
61. The idea patented was that of feeding energy produced from a solar-cell car-roof into the energy grid; the person who applied for the patent was, interestingly enough, a certain Mr Piëch, Director of Technology at Audi at the time.
62. At present, only Renault-Nissan have a deliverable vehicle; their Leaf is designed to operate within this kind of system. However, it remains to be seen whether other manufacturers will adopt this model.

REFERENCES

The following list contains all written sources for this work which are available in English. However, much of the research carried out on the subject of Volkswagen, as well as supporting information in other chapters, relies on sources only available in German. A note about these sources follows the English-language list.

Unless otherwise stated, all Internet addresses referenced were last verified between June 13 and July 29, 2011.

SOURCES AVAILABLE IN ENGLISH

ACEA, European Automobile Manufacturers' Association, "How much does your car emit?" (2011), available at http://www.acea.be/index.php/news/news_detail/how_much_does_your_car_emit

autoblog.com, "Chademo suggests drinking green tea while recharging your electric car," March 15, 2010, available at http://green.autoblog.com/2010/03/15/chademo-suggests-drinking-green-tea-while-recharging-your-electr/

Automotive News, "Toyota China capacity boosted to 820,000 vehicles," May 21, 2010, available at http://autonews.gasgoo.com/auto-news/1015269/toyota-china-capacity-boosted-to-820–000-vehicles.html

Automotive News, "U.S. car and light-truck sales, June," July 1, 2010, available at http://www.autonews.com/apps/pbcs.dll/article?AID=/20100701/RETAIL01/100709995/1448

Automotive News, "Toyota sees Prius beating 2010 U.S. sales as supply grows" July 8, 2011, available at http://www.autonews.com/apps/pbcs.dll/article?AID=/20110708/OEM05/307089869/1186

Automotive News Europe, *Global Market Data Books* (2005–2008).

autotropolis.com, "Toyota Kicks Off a New Sales Battle," March 16, 2010, available at http://www.autotropolis.com/automobile-news-columns/krome-on-cars/toyota-kicks-off-a-new-sales-battle.html

BBC Online, "How hybrid power surprised the car industry," November 16, 2004, available at http://news.bbc.co.uk/2/hi/business/4015831.stm

Berger, Roland, "Global sales of electric vehicles will rise to ten million in 2020", *Powertrain 2020 – The Future Drives Electric* (2009), available at http://www.

rolandberger.com/media/press/releases/519-press_archive2009_sc_content/Global_sales_of_electric_vehicles.html

Bloomberg.com, "Toyota Rating Cut by Moody's on View of Weak Profits" April 22, 2011, available at: http://www.bloomberg.com/news/2010-04-22/toyota-rating-cut-one-level-by-moody-s-on-profit-outlook-recall-costs.html

Bloomberg.com, "Yen Heading to New High as No Intervention in Sight With Fed Rates on Hold" June 20, 2011, available at: http://www.bloomberg.com/news/2011-06-20/yen-heading-to-new-high-as-no-intervention-in-sight-with-fed-rates-on-hold.html

Bloomberg.com, "VW Sees Passat Picking Up Jetta's U.S. Success," July 13, 2011, available at http://www.bloomberg.com/news/2011-07-13/vw-sees-passat-picking-up-where-jetta-is-in-u-s-market-cars.html

British Petroleum, *BP Statistical Review of World Energy* (2009).

Canis, Bill and Yacobucci, Brent D., *The U.S. Motor Vehicle Industry: Confronting a New Dynamic in the Global Economy* (Congressional Research Service, 2010).

China Automotive Research Center, *China Automobile Industry Yearbook* (2008).

China Daily, "Toyota feels the heat in China," March 9, 2010, available at http://www.chinadaily.com.cn/cndy/2010-03/09/content_9557732.htm

China Daily, "VW Group and China: A friendship of giants" (by Martin Winterkorn), April 24, 2010, available at http://www.chinadaily.com.cn/cndy/2010-04/24/content_9770052.htm

Central Intelligence Agency (CIA), *The World Factbook* (2010), available at http://www.cia.gov/library/publications/the-world-factbook/index.html

CNN Money, "Toyota's woes: Lower sales, ratings cut," June 28, 2011, http://money.cnn.com/2011/06/28/news/international/toyota_earthquake/index.htm (contains video interview with Akio Toyoda).

Crutzen, Paul J., Mosier, A. R., Smith, K. A. and Winiwarter, K. A. "N20 release from agro-biofuel production negates global warming reduction by replacing fossil fuels," *Atmospheric Chemistry and Physics Discussions*, 2007, Volume 7, July, 11191–205, available at http://www.atmos-chem-phys-discuss.net/7/11191/2007/acpd-7-11191-2007.html

Dyer, J.H. and Ouchi, W.G., "Japanese-style business partnerships: Giving companies a competitive edge," *Sloan Management Review*, Volume 35, Autumn 1993, pp. 51–63

Edmunds.com, "Edmunds.com Tracks Car Buyer Loyalty: Toyota Still Leads, but Loses Ground," March 15, 2010, available at http://www.edmunds.com/about/press/edmundscom-tracks-car-buyer-loyalty-toyota-still-leads-but-loses-ground.html

Edmunds.com, "One Year Later, Lingering Safety Concerns Not All that Ails Toyota, http://www.autoobserver.com/2011/01/one-year-later-lingering-safety-concerns-not-all-that-ails-toyota.html

egm CarTech, "Ford completes debt restructuring, reduces debt by $9.9 billion," April 6, 2009, available at http://www.egmcartech.com/2009/04/06/ford-completes-debt-restructuring-reduces-debt-by-99-billion/

Enerdata, *The Impact of Lower Oil Consumption in Europe on World Oil Prices* (Grenoble, 2009).

European Federation of Transport and Environment, *Reducing C02 Emissions from New Cars: A Study of Major Car Manufacturers' Progress in 2008*, available at http://www.transportenvironment.org

Fortune500 global, "GM's stock plans are a record," available at http://www.fortune500global.com/news/gms-stocks-plans-are-a-record/

Gartner Inc., "Hype Cycle for emerging technologies," available at http://www.gartner.com/pages/story.php.id.8795.s.8.jsp

GM Inside News, "GM to keep 900 more dealerships than planned," available at http://www.gminsidersnews.com/forums/f12/gm-keep-900-more-dealerships-than-planned-92611/

greenmomentum.com, "Tepco joins Japanese automakers to create global EV charging standard," March 15, 2010, available at http://www.greenmomentum.com/wb3/wb/gm/gm_content?id_content=5324

Heritage Foundation, "Auto Bailout Ignores Excessive Labor Costs," November 19, 2008, available at http://www.heritage.org/Research/Reports/2008/11/Auto-Bailout-Ignores-Excessive-Labor-Costs (by James Sherk, 2008).

International Energy Agency, *Oil Market Report 2009*, available at http://www.oilmarketreport.org/

Itazaki, Hideshi, *The Prius that Shook the World: How Toyota Developed the World's First Mass-Production Hybrid Vehicle* (transl. Albert Yamada & Masako Ishikawa, Nikkan Kogyo Shimbun Ltd., Tokyo, 1999).

J.D. Power and Associates, 2011 *Vehicle Dependability Study.*

just-auto.com, "Japan: Toyota passes 2.5m sales of hybrid vehicles," June 28, 2010, available at http://www.just-auto.com/news/toyota-passes-2.5m-sales-of-hybrid-vehicles_id104893.aspx?lk=dm

Kovarik, Bill, "Henry Ford, Charles Kettering and the 'Fuel of the Future'," available at http://www.runet.edu/~wkovarik/papers/fuel.htm

Lawson, John and Watkins, Philipp, "New Volkswagen: VW 'ex al' – A look at the company post merger," Citi Report, 2010

Liker, Jeffrey, *The Toyota Way: Fourteen Management Principles from the World's Greatest Manufacturer* (McGraw-Hill, New York, 2004)

MarkLines, Automotive Information Platform, *World Vehicle Sales Report* (2010)

McKinsey & Company, "Understanding China's Passenger Car Market," June 7, 2010, available at http://www.businessgreen.com/business-green/news/2239795/china-e6-electric-car-trying (presentation given on 06/07/2010)

msnbc.com, "Feds sign off on Toyota gas pedal remedy," January 31, 2010, available at http://www.msnbc.msn.com/id/35143209/ns/business-autos/

Next big future, "China has 100 million electric bikes," July 15, 2009, available at http://nextbigfuture.com/2009/07/china-has-100-million-electric-bikes.html

Osono, Emi, Norihiko Shimizu, and Hirotaka Takeuchi, *Extreme Toyota: Radical Contradictions that Drive Success at the World's Best Manufacturer* (John Wiley & Sons, Hoboken, NJ, 2008)

PriceWaterhouseCoopers, *Autofacts*, Quarterly forecast update.

Reuters, "GM falls just shy of Toyota in 2010 sales ranking," January 24, 2011, available at: http://www.reuters.com/article/2011/01/24/retire-us-toyota-gm-idUSTRE70N13K20110124

Reuters, "Ford profits despite economy, rising costs," July 26, 2011, available at: http://www.reuters.com/article/2011/07/26/us-ford-idUSTRE76P20Y20110726

Reuters, "GM Trucks Sitting on Lots as Americans Choose Fuel Efficient Cars," July 25, 2011, available at: http://www.reuters.com/article/2011/07/25/idUS79378774920110725

Rodriguez, Ann and Page, Chris, "A comparison of Toyota and Honda Hybrid Vehicle Marketing Strategies," Rocky Mountain Institute (2004), available at http://www.solsustainability.org/documents/cultivatingmarkets/A comparison of hybrid vehicle marketing strategies.pdf

Romney, Mitt, "Let Detroit Go Bankrupt," *New York Times*, November 19, 2008, available at http://www.nytimes.com/2008/11/19/opinion/19romney.html

Sato, Masaaki, *The Toyota Leaders: An Executive Guide* (Vertical Verlag, New York, 2008).

Simms, James, "Heard on the street: Toyota's next road test: used cars," *Wall Street Journal*, February 22, 2010.

Spear, Steven J., Working Paper 02–043, "Just-in-Time Practice at Toyota: Rules – In-use for Building Self-Diagnostic, Adaptive Work Systems" (2002).

Stewart, Thomas A., and Rama, Anand P., "Lessons from Toyota's long drive," *Harvard Business Review*, 7, July, 2007.

The Auto Channel, "Ford Motor Company Sets New Full Year US Sales Record," January 4, 2001, available at http://www.theautochannel.com/news/press/date/20010103/press033466.html

The Truth About Cars, "Chrysler Stops Building Cars," December 17, 2008, available at http://www.thetruthaboutcars.com/chrysler-adjusts-production/

Toyoda, Eiji, *Toyota: Fifty Years in Motion* (Kodansha International, Tokyo, 1987).

Toyota Motor Corporation, *Toyota: The History of the First 50 Years* (Toyota City, 1988)

Toyota Motor Corporation, Annual Reports 2000–2011, available at http://www.toyota.co.jp/en/ir/financial_results/2011/index.html

Toyota Motor Corporation, Half-Year Results 2011, available at http://www2.toyota.co.jp/en/news/11/07/0725.html

Toyota Motor Corporation, Toyota Group's New Corporate Structure Aimed at Strengthening Production in Japan, available at http://www2.toyota.co.jp/en/news/11/07/0713.htm

United Nations, Department of Economic and Social Affairs, Population Division, *Urban and Rural Areas* (2007), available at http://www.un.org/esa/population/publications/wup2007/2007urban_rural.htm

US Energy Information Administration, *Annual Energy Outlook 2010*.

Volkswagen Aktiengesellschaft, Press releases and publications 2000–2010.

Volkswagen Aktiengesellschaft, Annual Reports 2000–2009, available at http://www.volkswagenag.com/vwag/vwcorp/content/en/homepage.html

Womack, James P., Daniel T. Jones, and Daniel Roos, *The Machine that Changed the World* (Harper Paperbacks, New York, 1991).

Xinhua News, "Beijing to Revive Bikes for Smooth Traffic, Clear Sky," January 24, 2010, available at http://www.china.org.cn/wap/2010-01/24/content_19301904.htm

Xinhua News, "China aims to be world pacemaker of new energy auto production," April 15, 2009, available at http://www.china.org.cn/business/2009-04/15/content_17612829.htm

Yahoo Autos, "Top Cash for Clunkers Trade-Ins and New Cars," available at http://autos.yahoo.com/articles/autos_content_landing_pages/1036/top-cash-for-clunkers-trade-ins-and-new-cars/

A NOTE ON GERMAN LANGUAGE SOURCES

This book was conceived and written in German and takes as one of its central topics the Volkswagen Group, based in Wolfsburg in Northern Germany. As such, a large number of the original sources referenced are only available in German.

A wide range of articles from the quality German press were cited, predominantly from *Handelsblatt*, Germany's leading financial daily (http://www.handelsblatt.com). Other frequently referenced media sources include the *Financial Times Deutschland* (http://www.ftd.de), the weekly magazines *Der Spiegel* (http://www.spiegel.de) and *Wirtschafstwoche* (http://www.wiwo.de), as well as German-language reports produced by Reuters.

Two smaller media sources on which the book draws are *StadtAnsichten*, based in VW's home city Wolfsburg and styling itself as "the magazine for the car city" (http://www.autostadt.de), and *Wolfsburger Allgemeine Zeitung*, also based there. In the former, an interview with Ferdinand Piëch published in October 2008 was frequently referenced; the latter published an interview with Bernd Osterloh in December 2009.

Two other sources are only available in German and yet worthy of note here due to their effect on the book. The authors frequently cited research on innovation undertaken within car companies in the years 2009 and 2010 by the Center of Automotive Management in Bergisch Gladbach, near Cologne; of no small importance to the work on Toyota was a 2007 publication by Ulrich Jürgens whose title is best translated as "How has Toyota stayed so strong for so long?", Wissenschaftszentrum für Sozialforschung, Berlin, 2007.

INDEX

A1 (Audi), 23
A123 Systems, 60
A8 (Audi), 35
acceptance test drives, 37
Acura, 158
Advanced Traction Battery Systems (ATBS), 60
advertising, 44–5
affordability, 22
Akerson, Daniel, 9
Alfa Romeo, 77–8
alternative energy sources, 15, 18, 19, 48
alternative energy vehicles, 59–60, 193–209, 217–18
  see also electric vehicles (EV); hybrid vehicles
Amarok, 65
American Axle, 8
American consumer tastes, 188–9
andon, 137, 139, 143, 150–1
Aston Martin, 176
Audi, 17, 23, 40, 43, 46, 161, 221
  branding of, 76
  culture of, 74–5
  green technologies, 92
  modular toolkits, 70
  profitability, 78
  technological innovation by, 35–6
Audi Quattro, 35
Auris (Toyota), 97, 98
Auto 5000 GmbHH, 105
auto engines, 15, 35, 52, 89, 90–104, 110
auto industry
  2010 sales and revenue, 13
  Chinese, 58–61
  global sales forecast, 14
  rebalancing of global, 14–15
automation, 154–5
Autostadt, 38
Auto-Union, 39
Aygo, 24, 224

batteries, 100–3, 201–2, 205, 216, 243–6, 250–1
battery-driven vehicles, 19, 48, 59, 60, 201–2
Beetle (VW), 27, 41, 43
Beijing Motor Show, 47–8
Bentley, 40, 76
Bernhard, Wolfgang, 219
Better Place, 249–51
Big Three automakers, 7–9, 186–8
biodiesel, 232–4
bioethanol, 233–4
biofuels, 231–6
biogas, 236
biomass, 234–6
Blue Motion technologies, 52, 91–2, 215
BlueSport, 77
BMW, 17, 39, 70, 158, 159, 161
branding, 17
  by Toyota, 156–7
  by Volkswagen, 44–5, 75–81, 89–90
Bratislava plant, 73
Brazil, 65, 73, 85, 155, 162, 224
Browning, Jonathan, 43
Buffett, Warren, 61
Bugatti, 40
Bugatti Veyron, 90
Build-Your-Dreams (BYD), 61–3, 101

Bush, George W., 6
business strategy
  of Toyota, 172–6
  of Volkswagen, 26–7, 33–41

California, emissions regulations in, 15–16, 238
California Air Resources Board (CARB), 16
California Clean Air Act, 238
Camry (Toyota), 43
capital costs, 36
carbon emissions, 15–18, 90–1, 98–9, 199, 201
car segments
  development of, 21
  transformation of, 20
Cash for Clunkers program, 11
Chademo initiative, 204
Chattanooga, Tennessee, 2, 3, 42, 43, 46, 73, 83, 106, 221
Chavez, Hugo, 165
Chevrolet Cruze, 10–11
Chevy Volt, 10, 20
China
  auto manufacturing in, 58–61
  auto market in, 13–14, 34, 47–58
  car sales in, 53
  economic boom in, 47–58
  GM sales in, 10
  low-cost models in, 22
  manufacturing growth in, 58–61
  monetary policy, 51
  sales forecasts for, 50
  Toyota and, 63–5, 133, 155
  US auto makers in, 13
  Volkswagen in, 34–5, 42, 47–58, 73, 74, 79, 85–6, 113, 217–18, 223–5
  VW in, 25
China Automotive Technology and Research Center (CATARC), 63
Chrysler, 7–9, 121
  comeback of, 12–13
  government bailout of, 6
class action lawsuits, 180, 186
Clean Air Act (CAA), 15
clean air legislation, 15–16
Clean Vehicle Rebate Project, 16
climate change, 15
CO2 emissions, 15–18
Collins, Jim, 132–3
commercial vehicles, 31–2, 41
compact cars, 8–11, 24, 38–9, 174–5, 224
compact car wars, 186–7
competitive environment, 14–15
components, standardization of, 68–74, 147, 153, 156, 222
concept cars, 18
continuous improvement, 81, 137–8, 140–7
Continuous Improvement Process (CIP), 37
Corolla (Toyota), 4, 24, 157, 188
Corporate Average Fuel Economy, 15
corporate decline, 132–3
corporate social responsibility, 196
corporate values, 226–7
cost control, 36
Creddisa, 159
customer satisfaction, 81
Czech Republic, 35, 39

Dacia, 84
Dacia Logan, 22
Daihatsu, 167
Daimler, 17, 23, 31, 39, 62, 158, 159, 161, 176
dealerships, 45–6, 67–8, 187
Delphi, 8
Demel, Herbert, 113
de'Silva, Walter, 33
Detroit Motor Show, 17, 43, 196, 205
direct injection diesel engine, 35, 89
direct shift gearboxes (DSG), 89
DSG transmission, 52
Ducati, 77
Dürrheimer, Wolfgang, 76

economies of scale, 35
electric bikes, 60, 246
electric infrastructure, in China, 60
electricity pumps, 248–50
electric vehicles (EV), 10, 19, 23, 48–9, 59, 60, 62–3, 217–18, 237–54
  batteries, 239–46
  batteries for, 100–2
  consumer demand for, 251–2

electric vehicles – *continued*
  cost of ownership, 93–7
  early, 237–9
  environmental advantages of, 239–40
  infrastructure for, 248–51, 252–4
  micro-cars, 246–7
  plans for, 102–3
  product strategy for, 99–100
  profitability, 93–7
  progress on, 247–8
  research and development of, 102
  by Toyota, 203–7
  workings of, 240–3
Emergency Profit Improvement Program, 144
emerging markets, 34–5, 39, 47
  low-cost models for, 84–9
  Toyota in, 65, 134, 162–4, 222–5
  at Volkswagen, 220–1
  Volkswagen in, 65–6, 84–9, 222–5
employees, 34, 72–3
  role of, at Volkswagen, 104–13
  at Toyota, 151–2, 221
employment, green technologies and, 19
engineering hours per vehicle (EHPV), 82
Environmental Protection Agency (EPA), 15
Etios, 162, 164–5, 224
euro-dollar exchange rate, 46
Europe
  car market in, 22, 72, 154, 197–8
  emissions regulations in, 17–18
EV1, 238
exchange rates, 128
Exeo, 76

Ferrostaal, 32
Fiat Group, 12, 65
financial crisis, 5, 9, 11, 28, 123, 125, 126, 128, 130
First Auto Works (FAW), 50
Ford, Henry, 231
Ford Focus, 11
Ford Motor Company, 4, 7–9, 11–12, 121, 135, 136, 225
foreign factories, 72–5, 127–8, 149–58
Formula 1 racing, 132
Fremont plant, 132
fuel cells, 207–9, 236–7
fuel efficiency standards, 15–18
fuel-efficient cars, 10–11, 42–3, 90–104
  see also hybrid vehicles

gasoline prices, 8–9, 21
Geely Holding Group, 13–14, 43
General Motors, 7–9, 46, 50, 121, 125, 132, 138, 175, 180, 206
  China and, 13
  comeback of, 9, 210–12, 225
  competitors of, 7
  government bailout of, 6–7
  new models by, 10–11
  restructuring of, 7
  stock price, 10
German auto industry, 3–4
German company law, 28, 32
Germany, participative management in, 104–5
globalization, 222–5
Golf (VW), 24, 37, 38, 65, 77, 78, 89, 97, 98, 103, 155, 188
government regulations, on vehicle emissions, 15–18
Great East Japan Earthquake, 4–5, 46–7, 122, 124, 127, 150, 209, 211
Great Recession, 5, 9
green cars, 17, 20–2, 27, 48, 62–3, 193–209
  see also electric vehicles (EV); hybrid vehicles
greenhouse gas emissions, 15–18, 90–1, 98–9, 199, 201
green technologies, 18–22, 59–61, 90–104, 217–18
growth markets, 34–5
  see also emerging markets
Günak, Murat, 115

Hahn, Carl, 41, 56
Härter, Holger, 28, 29
Hartz, Peter, 105
Heizmann, Jochem, 31, 80, 83
Highlander (Toyota), 127
Hino, 167
Honda, 158, 162, 186, 187, 200, 216
Horch, August, 81

Hummer, 6
hybrid effect, 96
hybrid vehicles, 17, 19, 21, 24, 48, 49, 61, 63, 64, 196–203, 214–17
  affordable, 97–9
  cost of ownership, 93–7
  profitability of, 93–7
  from VW, 26–7, 90–104
hydrogen fuel cells, 207–9, 236–7
Hyundai, 22–3, 49, 79, 87, 164, 213

Inaba, Yoshimi, 122
India, 73, 85, 224
  low-cost models in, 22
  Toyota in, 162, 164–5
  Volkswagen in, 35, 39, 65, 66, 86–7
Infiniti, 159
in-house production, 82–3, 101
Insight (Honda), 216
insurance liabilities, of Big Three, 8
International Multipurpose Vehicle (IMV) models, 163–4
International Works Council, 107, 108
itaku, 167

Jacoby, Stefan, 43
Japan
  Great East Japan Earthquake, 4–5, 46–7, 122, 124, 127, 150, 209, 211
  R&D in, 194–6
  role of government in economy in, 194–5
Japanese auto industry
  Great East Japan Earthquake and, 5
  Toyota, 4–5
Jazz (Honda), 200
Jetta (VW), 27, 43, 216
joint ventures, 39, 50, 54–7, 60, 64, 66, 132, 138, 175, 223–4
just-in-time production, 136–7

kairetsu, 165, 168, 170–2, 219
kaizen, 81, 140–7, 151, 168–9
kanban, 136–7
Kanto Auto Works, 167
K-cars, 87
Kia Motors, 22, 79
Klinger, Christian, 66, 67, 68
knowledge transfer, 55–6, 58
Kolin factory, 153
Kortüm, Franz-Josef, 113
Krebs, Rudolf, 49, 96, 102

labor costs, 74, 219
labor unions, *see* unions
LaHood, Ray, 3, 179, 181
Lamborghini, 40, 77
Lancia, 12
Latin America, 65
Lavida, 52
layoffs, 8
Leaf (Nissan), 10, 206
lean manufacturing, 135–52, 154–5, 220
Lexus, 4, 17, 45, 64, 81, 156, 158–61
lithium-ion batteries, 60, 61, 100–2, 205, 243–6
lobbying, 45
local factories, 73
localization, 44, 57, 73, 74
López de Arriortúa, José Ignacio, 108–9
low-cost models, 22–3, 58, 84–9, 162–4, 222–5
Lupo (VW), 98
luxury brands, 17, 22–4, 64, 76, 89–90, 158–61

Macht, Michael, 83
MAN, 31–2, 40, 77
Marchionne, Sergio, 12
Maruti Suzuki Alto, 87
Mazda, 197
Mercedes, 17, 39, 89, 160
mergers and acquisitions, 27–33, 39, 40–1, 87–8
methane, 236
micro-cars, 22, 24, 87, 205, 224, 246–7
Ministry of Economy, Trade and Industry (METI), 194
Mitsubishi, 23, 162, 175, 206, 217
Model T, 231
modernization, 82
modular longitudinal toolkit (MLB), 70
modular toolkits, 35–6, 68–72, 81, 85, 213, 216, 219, 223
Modular Transverse Toolkit (MQB), 70–1
muda, 135, 155
Muir, James, 76

Mulally, Alan, 11
Müller, Matthias, 103, 114
Museum of Modern Art, 1

Nanjing Auto, 59
Nano, 84, 86–7, 246
Neumann, Horst, 34, 72, 104
Neumann, Karl-Thomas, 49, 56, 96, 117–19
new European Driving Cycle (NEDC), 36
niche products, 18
Nissan, 136, 158–9, 162, 170, 186, 187, 206
Nummi, 138, 175–6

Obama, Barack, 6, 15, 18
Ohno, Taiichi, 136
oil, 14–15
oil supplies, 228–31
Olympic Games (2008), 56
Opel/Vauxha, 7
operating profit, 36–9, 78, 144, 192
original equipment manufacturers (OEM), 38, 47–8
Osterloh, Bernd, 66, 106–8, 110, 116
outsourcing, 8, 83, 101, 166
Ozawa, Satoshi, 64

Paefgen, Franz-Josef, 113
participative management, 104–13, 219
parts supply, 44
  see also suppliers
Passat (VW), 3, 43
pension costs, 8
Peters, Jürgen, 105
Peugeot, 23, 175, 217
Phaeton (VW), 80, 86, 97
Piëch, Ferdinand, 27, 31–3, 35–6, 74, 75, 77–8, 88, 92, 103, 105–6, 112–16
Pischetsrieder, Bernd, 77, 114
plug-in hybrid vehicles (PHVs), 97, 201–3, 215–16, 247–8
politics, Volkswagen and, 104–13
Polo (VW), 79, 80, 86–7, 215
Pontiac, 6
population growth, 228–31
Porsche AG, 27–32, 36, 41, 57, 67–8, 71, 90
Porsche Automobil Holding SE, 27, 28–9
Porsche Holding Salzburg, 68
Posche, Ferdinand, 27
Pötsch, Hans Dieter, 26, 29
powertrain technology, 18, 19, 24, 155, 157, 216
premium car segment, 17, 22–4, 220–1
Prius (Toyota), 4, 27, 61, 64, 90, 97–9, 127, 145, 149, 180, 196–203, 215, 217
production
  see also Toyota Production System (TPS)
  foreign factories, 72–5, 127–8, 149–58
  in-house, 38, 82–3, 101
  at Volkswagen, 81–4
productivity, 73, 146, 219, 221
profitability, 25
  green technologies and, 20–2
  of hybrids and EVs, 93–7
  Toyota, 126, 134, 144, 155–6, 192
  Volkswagen, 33, 36–9, 78, 218–22
public relations, 17, 176, 180, 192
pull system, 136–7

Qatar, 28
quality issues, 72, 120–4, 129

Rabbit (VW), 2
rebates, 44
recalls, 4, 63–4, 71–2, 81, 120–4, 129, 176–90, 212–13
Renault, 23, 206
Renault-Nissan, 176, 206, 217
research and development
  in green technologies, 18–19, 59–60, 102, 194
  investment in, 15
  in Japan, 194–6
  by Toyota, 129
robotics, 194, 195–6
Romney, Mitt, 8
"runaway Toyotas", 176–90
Russia, 23, 35, 65, 66, 72, 85, 224

Saab, 6
sales and marketing, by Volkswagen, 66–8
Sanyo, 205

Sato, Masaaki, 130
Satoshi, Kamata, 151
Saturn, 6
Scania, 31, 39, 40
Schmidt, Werner, 74
SEAT, 35, 39, 40, 76
Sensei, 82
SGL Carbon, 92
Shanghai Automotive Industrial Corp. (SAIC), 50, 54, 59
Shanghai Volkswagen Automotive Company, 34, 55
Skoda, 35, 39, 40, 49, 75–6, 84, 85, 89–90, 221
Slater, Rodney, 181, 189–90
small cars, *see* compact cars
Smith, Graham, 208
soccer clubs, 44–5
South America, 65
Spain, 35, 39
standardization, 68–74, 147, 153, 156, 222
Super Bowl commercial, 45
Super-Efficient Vehicle (SEV), 92–3
suppliers, 38, 44, 71, 73, 103, 130–1, 165–72, 179
supply chain management, 165–72
SUVs, 9, 127
Suzuki, 24, 39, 41, 66, 84, 87–8, 162, 223
Suzuki, Osamu, 1, 88
synergies, 39–41

Takaoka, 154–5
Tata Nano, 84, 86–7, 246
tax credits, for electric cars, 10
technological innovation, 15, 18–22, 24, 172, 217
  by Audi, 35–6
  green technologies, 59–61, 90–104
  by Toyota, 155, 156
  by Volkswagen, 38, 52
technology transfer, 55–6, 64, 88, 174–5
Tesla, 17, 158, 175–6, 193, 206–7, 218, 243
TFSI, 89
Tiguan (VW), 80
Touareg (VW), 24, 80, 90, 215
Toyoda, Akio, 4, 5, 19, 63–4, 120–3, 128, 131–4, 158, 178, 179, 190–3
Toyoda, Eiji, 136
Toyoda, Sakichi, 130, 135
Toyota, 4–5, 63
  affiliated companies, 166–72
  branding, 156–7
  business strategy, 126
  in China, 155
  China and, 13, 63–5, 133
  competition between VW and, 46–7, 210–27
  corporate culture, 151–2, 175
  corporate values, 226–7
  credit rating of, 124
  crises facing, 120–31, 176–90, 211
  design philosophy, 157–8
  electric vehicles (EV), 203–7, 217–18
  in emerging markets, 65, 134, 162–4, 222–5
  employee relations, 153
  factories of, 147–58
  fuel efficient models, 17, 18
  global strategy, 25, 152–8
  green cars and, 193–209
  growth of, 129–31
  handling of crisis by, 178–81, 190–3, 212–13
  history of, 134–9
  hybrid vehicles, 17, 21, 24, 196–203, 215, 216
  hydrogen fuel cells and, 207–9
  image of, 121, 123–6, 179–80
  Japanese identity of, 152–8
  joint ventures, 175
  Kaizen, 81
  labor conflict at, 124–5
  long-term strategy of, 172–6
  luxury brands, 158–61
  management style, 130, 190–1
  micro-cars, 23–4
  new beginning for, 131–4
  ownership structure, 172–4
  philosophy of, 25, 144–5, 220
  production system, 135–58, 219, 220
  profitability, 126, 134, 144, 155–6, 192, 218–19
  quality issues, 4
  R&D by, 19
  recall disaster, 63–4, 71–2, 120–4, 129, 176–90, 212–13

Toyota – *continued*
supplier management, 165–72
technological innovation, 155, 156
US factories, 127–8
US market and, 44, 45, 126–7, 186–9
Toyota City, 145–6, 149, 152
Toyota Industries, 134–5
Toyota Production System (TPS), 131, 134, 135–52
"Toyota Shock", 123–31
Toyota Way, 134–5
TPS, *see* Toyota Production System (TPS)
trade disputes, 186–8
truck market, 31–2
TSI engine, 52

Uchiyamada, Takeshi, 63
unions, 8, 104–5, 124–5, 151–2, 221
Up (VW), 85, 87, 97, 100
US auto industry
Big Three, 7–9, 186–8
China and, 13
financial crisis and, 5–6, 11
government bailout of, 6–7, 9
layoffs, 8
US auto market, 73
financial crisis and, 9
Lexus and, 160–1
Toyota and, 4–5, 25, 44, 45, 126–7, 186–9
Volkswagen and, 2–4, 25, 41–6, 225
US Congress, 120–3
US dollar, 128

Vahland, Winfried, 73, 76, 85
vehicle dependability, 188–90
vehicle emissions, regulations on, 15–18
vehicle-to-grid technology, 249
Vfl Wolfsburg, 38
Visteon, 8
visualization, 141–2
Volkert, Klaus, 105
Volkswagen, 17
brands, 40–1
brand strategy, 44–5, 75–81, 89–90
business strategy, 26–7, 33–41
in China, 74, 79
China and, 13, 34–5, 42, 47–58, 85–6, 113, 217–18, 223–5
commercial vehicles, 31–2, 40
competition between Toyota and, 46–7, 210–27
competitors of, 4, 5, 46–7
corporate values, 226–7
electric vehicles (EV), 99–100, 102–3, 241
in emerging markets, 47, 65–6, 84–9, 222–5
engine types, 90–104, 110
factories of, 72–4
fuel efficiency technology and, 24
fuel efficient models, 17, 18
global positioning strategy, 25
green technologies and, 48–9
growth of, 72–5, 213–14, 221
high standards of, 36–7
hybrid vehicles, 26–7, 90–104, 214–17
in India, 86–7
in-house production, 38
low-cost models, 84–9, 162
luxury brands, 89–90
mergers and acquisitions, 27–33, 39–41, 87–8
modular toolkits, 35–6, 68–72, 213, 216, 219, 223
MoMA and, 1
operating profit, 33
participative management at, 104–13, 219
philosophy of, 25, 220, 222
politics at, 104–13
premium car segment, 220–1
production system, 81–4
profitability, 36–9, 78, 218–22
R&D by, 19
sales push by, 66–8
suppliers, 171
synergies, 39–41
technological innovation, 38, 52
Toyota and, 180
US factories, 2–4, 41, 42, 43, 46
US market and, 41–6, 225
"Volkswagen Way", 37, 81–4
Volt (GM), 206
Volvo, 13–14, 43

Wagoner, Rick, 6
Wang, Chuanfu, 61, 62

Watanabe, Katsuaki, 127, 130, 146, 195
Wiedeking, Wendelin, 28, 29, 78, 114
Winterkorn, Martin, 1–3, 26–7, 29, 32–4, 36–7, 39, 47, 48, 50, 72, 74–5, 77–80, 97, 102, 107, 110–11, 114–16, 131
Wittig, Detlef, 88
world car, 223

XL1 SEV, 92–3

Yamashina, Tadashi, 132
yen, 128

zero emissions, 17, 48–9, 205, 207